P9-CKT-290

BOURNE, Geoffrey Howard and Maury Cohen. The gentle giants: the gorilla story. Putnam, 1975. 319p ill bibl index 75-25753. 12.50. ISBN 0-399-11528-5. C.I.P.

CHOICE   OCT. '76
Biology

An accurate, nontechnical, and sensational narrative of gorillas that ranges from the time of Du Chaillu's killing of wild gorillas in the late 1800s to present-day studies in the natural habitat. Much of the text and most of the 85 photos are devoted to gorillas, chimpanzees, and other monsters in movies, show business, circuses, zoos, and the Yerkes Primate Center. There is a list of all the American movies in which gorillas have appeared. Only 20 pages are devoted to the recent field studies of George Schaller and Dian Fossey. This is a hodge-podge of anecdotal material including anatomical notes on ape body size, attempts to teach chimpanzees to talk, descriptions of sexual behavior, and a chapter on the hunt for the yeti. *The great apes* by R. M. and A. W. Yerkes (1929) is a more thorough historical account, and the original reports by G. B. Schaller in *The mountain gorilla* (1963) and *The year of the gorilla* (CHOICE, Jul.–Aug. 1964) are more authoritative accounts of a recent field study. Written by the director of the Yerkes Primate Center in collaboration with a movie producer/writer, the present work is similar in style to Bourne's previous book, *The ape people* (CHOICE, Jun. 1971). Well indexed; sketchy bibliography.

QL
737
P96
B77

The Gentle Giants
The Gorilla Story

DR. GEOFFREY HOWARD BOURNE is the director of the Yerkes Regional Primate Research Center, Emory University, Atlanta. He is the author of numerous books and articles, including *The Ape People* and *Primate Odyssey*.

MAURY COHEN is an award-winning producer and writer for motion pictures and television, whose credits include such TV successes as *You Asked for It* and *The Ernie Kovacs Show*. His special interest in animals directed him to a three-year association with the San Diego Zoo.

The Gentle Giants
The Gorilla Story

Geoffrey H. Bourne
and Maury Cohen

G. P. PUTNAM'S SONS
NEW YORK

LIBRARY

|MAR 2 4 1976

UNIVERSITY OF THE PACIFIC

310969

Copyright © 1975 by Geoffrey H. Bourne and Maury Cohen

*All rights reserved. This book, or parts thereof, must not
be reproduced in any form without permission. Published simultaneously
in Canada by Longman Canada Limited, Toronto.*

SBN: 399-11528-5

Library of Congress Cataloging in Publication Data

Bourne, Geoffrey Howard, 1909–
    The gentle giants.

    Bibliography: p. 303
    Includes index.
    1. Gorillas. I. Cohen, Maury, joint author. II. Title.
QL737.P96B67   1975   599'.884   75-25753

PRINTED IN THE UNITED STATES OF AMERICA

Preface and Acknowledgments

THERE is something about a gorilla that produces fear in the uninitiated, mainly because of the reputation built up for this animal by the early explorers and hunters. Perhaps there is also some racial memory of a dark, frightening creature in the woods that lingers in the depths of man's subconscious mind and helps make the gorilla more of a frightening apparition. Superimposed on all this, the movie industry, in its earliest films, depicted the gorilla as a monster and built up an image of enormous, sinister strength. The poor gorilla, who, left alone with his family, is a benign and gentle creature with no serious enemies to fear and content only to live in peace, has had to live with this image. He has been shot, captured, and exhibited so that humans can brag about how brave they were to kill him or capture him or just stand back in fear and awe and look at him in a cage.

The Yerkes Primate Center has fifteen gorillas, all of which arrived as lovable little babies who played in the nursery and fed from bottles containing human baby milk formula, and it was hard to think of them as violent or as frightening or even as very strong when they were at this early stage. There is no doubt, however, about their strength later when they have begun to grow, but to the Yerkes staff they have never become objects of fear. Their reactions are reasonably predictable, they have pleasant relations with the staff, many aspects of their personality are reminiscent of humans, and the staff enjoys working with them and making friends with them.

Within the pages of this book we have tried to tell the story of the gorilla, from the accounts of the early hunters, as he is seen through the eyes of those who went to study him in the wild and those who have kept him in captivity and studied him in the laboratory. He emerges as a gentle vegetarian, and perhaps those who read about him in these pages will learn a little how to understand and to sympathize with this wonderful and noble nonhuman primate, so close to ourselves yet separated by an evolutionary gulf which will make him forever just our cousin.

There are many people who have helped make this work possible—all the writers of books and scientific papers about gorillas which we have had to study to produce our book. We cannot acknowledge everywhere in the book where they have helped us, but we have listed more than one hundred persons and their publications in the Bibliography.

Many people have also been helpful to us in providing illustrations for this book, and we are most grateful to them for their cooperation. We want particularly to acknowledge the help provided by Bob Burns who was most helpful in the acquisition of movie pictures, in giving us intriguing information on what it is like to be a movie gorilla and who has helped the book in many ways.

The typing of the manuscript of a book is no light task, and the two secretaries of G.H.B., Gwen Cook and Linda Alderson, slaved over the manuscript four times before it was judged ready to go to the publisher. Frank Kiernan, Yerkes photographer, provided valuable photographic assistance, as he did with *The Ape People* and *Primate Odyssey*. William Targ, senior editor of Putnam's, was, as always, a sympathetic conductor into the world of book publishing.

GEOFFREY H. BOURNE

Yerkes Primate Research Center
Atlanta, Georgia

MAURY COHEN

Twentieth Century-Fox Film Corporation,
Los Angeles, California
May 28, 1975

CONTENTS

The Gentle Giants
The Gorilla Story

1

How to Create a Monster!

IT was the thickest part of the forest, somewhere near the Sierra del Crystal. The party of three black men and one white forced its way through the undergrowth, sweat streaming down their faces in the steamy heat. Stinging insects helped to make their progress slow and painful, and thorns on many of the vines and bushes added to their misery. Yet, in spite of their discomforts, they did not despair. Soon they came on trees which bore a pulpy, pear-shaped fruit that the Africans called *Tondo*. It was a fruit that gorillas love, and it was a gorilla that the party was after—where there was *tondo*, there was bound to be a gorilla. Near some rapids where the water rushed with intimidating speed between enormous boulders, was a deserted village. Scattered among the ruins of the huts were broken pieces of sugarcane and whole canes which had been pulled up by the roots and thrown about. Some had been chewed, and there, imprinted in the soft ground, were the tracks of a gorilla. The party had hardly taken in the fact that they were closing in on their quarry when suddenly, on the opposite side of the clearing, the bushes parted and an immense black hairy form appeared. It paused momentarily as if to judge the size and distance of the hunting party; then, uttering a terrible howl of rage, which resounded in the forest like the rolling of distant thunder, the animal made straight at them.

When he first appeared, sixty feet away, the white man raised his rifle and took aim. The black men whispered, "Not time yet." The animal stopped his rush about halfway and looked at the party again. He then beat his breast with his gigantic arms and once more advanced, this time halting abruptly thirty or so feet away. Again the black men whispered, "Not yet." The white man had a shuddering thought: "What if our guns misfire or if we only wound the huge beast?" At this distance the hunters could see the gorilla's face—it was distorted with rage. They could hear him grinding his teeth in anger. They could see him spasmodically contracting the skin on his forehead so that his scalp, pushed into an enormous crest on the top of his head, moved rapidly backward and forward—a face so evil it suggested the devil himself. The animal roared again. He beat his breast with his immense hands, each beat sounding like the firing of a cannon. Then he began his final rush at the hunters. A black man warned again, "Don't fire too soon. If you don't kill him, he'll kill you." The gorilla's rush carried him to within a few feet of the gun, and the black men shouted, "Now!" The white hunter fired, and the animal fell with a tremendous thud. He twitched a few times as he lay there and then subsided into the relaxation of death. The proud hunter mopped his brow as his African companions congratulated him, and the four of them bent to examine the animal. He had appeared eight feet tall as he charged them. In death he measured only five feet six inches, but his great arms, when spread, covered seven feet two inches. His brawny chest measured fifty inches around; his big toe measured five and three-quarter inches in circumference. The animal's face was jet black, with a flattened nose; the great chest was almost devoid of hair. The skin over it was stretched tight and was parchmentlike in appearance. The back was covered in gray hair.

The black hunters told the white man that in the forest there were many gorillas larger than this one. Some of their tribe had speared and killed much bigger animals. They told of the devious ways of the gorilla—how he would hide himself in the lower branches of the forest trees and kill a human who passed along the trail below. One African spoke of how a party of gorillas was found in a field tying up sugarcane in bundles to carry it away. The animals were attacked by a group of Africans, but the gorillas routed and killed many of them, carrying off others as prisoners. In a few days the Africans returned home, but only after the gorillas had tortured them by tearing off their fingers and toes.

One day, recounted an African, two Mbondemo women were walking together through the forest when an immense gorilla stepped in front of them, grabbed one of the women, and carried her away, struggling and screaming. The captured woman was released by the gorilla after a few days and returned to her house. But this was not the case in other stories they told of African women captured, ravaged, or killed by the giant animals.

The black hunters recalled the names of many dead men in their tribe whose spirits were known to be dwelling in gorillas. One African told the following story: "Several seasons ago a man suddenly disappeared from my village after an angry quarrel. Sometime after, an Ashira of that village was out in the forest. He met a very large gorilla. That gorilla was the man who had disappeared; he had turned into a gorilla. He jumped on the poor Ashira and bit a piece out of his arm, then he let him go. Then the man came back with his bleeding arm. He told me this. I hope we shall not meet such gorillas."

The white hunter, not satisfied with his trophy, wanted to capture a gorilla alive, and he set out again, this time with the help of five African hunters. While walking stealthily through the forest tracking their prey, they suddenly heard a young gorilla crying for its mother. Holding their guns ready, the brave hunters crept cautiously toward the baby. Seeing the bushes moving ahead of them, they dropped to a crawl and edged noiselessly forward, then stopped suddenly—ahead of them was a rarely seen sight. A baby gorilla was sitting happily on the ground eating berries. A few feet away sat the mother, contentedly munching the same fruit. As soon as they saw this peaceful and happy sight, the hunters raised their guns and pulled their triggers. The mother fell on her face, blood gushing from her body in several places. Terrified by the noise of the guns, the baby rushed to its mother for protection. He clung to her desperately and buried his face in her fur. The hunters immediately rushed forward toward the two, shouting with joy. Terrorized by the sounds of the humans, the baby, now covered with its mother's blood, ran to a small tree and quickly climbed it. Once in its branches, the orphaned gorilla roared defiance at its mother's killers. The hunters then cut down the tree and, as it fell to the ground, quickly threw a net over what they later described as a "young monster." In the process of restraining it, however, we are happy to relate that one of the hunters received a severe bite on the hand and another had a piece of flesh bitten out of his leg.

What happened next is best described in the words of the white hunter:

The little brute, though very diminutive, and the merest baby in age, was astonishingly strong and by no means good tempered. [Would you expect him to be?] They found they could not lead him. He constantly rushed at them, showing fight, and manifesting a strong desire to take a piece, or several pieces, out of every one of their legs, which were his special objects of attack. So they were obliged to get a forked stick, in which his neck was inserted in such a way that he could not escape, and yet could be kept at a safe distance.

The excitement in the village was intense as the animal was lifted out of the canoe in which he had come down the river. He roared and bellowed, and looked around wildly, with his wicked little eyes, giving fair warning that if he could get any of us he would take his revenge.

They made a cage for him and his courage was undiminished. He rushed at anyone who came near him. The white hunter said, "He sat in his corner, looking wickedly out of his gray eyes, and I never saw a more morose or ill-tempered face than this little beast had. I do not believe gorillas ever smile."

After having escaped twice from his cage and being recaptured, the baby gorilla was chained to a stake. Ten days later he died.

This horrible tale of the killing of a mother gorilla by hunters or poachers in order to capture the baby for sale to menageries or zoos has been repeated *ad nauseam* over the last 125 years, and it still goes on in a more limited but nonetheless clandestine way.

The white hunter of these stories we have just recounted was Paul Du Chaillu, a Franco-American who had decided to learn for himself the truth of the stories about monsters that had been coming out of Africa. He first went to Africa around 1850 and stayed eight years. The first four years he spent in commercial activities with his father; the last four were spent in exploring and in hunting gorillas. Du Chaillu's claims were substantial:

I travelled—always on foot, and unaccompanied by other white men—about 8,000 miles. I shot, stuffed and brought home over 2,000 birds, of which more than 60 are new species, and I killed upwards of 1,000 quadrupeds, of which 200 were stuffed and brought home, with more than 80 skeletons. Not less than 20 of the quadrupeds are species hitherto unknown to science. I suffered fifty attacks of the African fever, taking, to

14

Paul Du Chaillu shoots his first gorilla. From Paul Du Chaillu, *Explorations and Adventures in Equatorial Africa*

cure myself, more than fourteen ounces of quinine. Of famine, long-continued exposures to the heavy tropical rains, and attacks of ferocious ants and venomous flies, it is not worth while to speak.

When Du Chaillu returned to America, he was given a great reception. People shuddered at his stories of the black hairy monsters, the gorillas, he had seen and killed in Africa.

The Royal Geographical Society in England invited him to visit England to give a lecture before the fellows and members of the society. He was given a hero's welcome at first, but later, zoologists began to doubt his claims. The specimens he had exhibited at the Geographical Society which he represented as new species were shown not to be new species at all. His emotional description of his encounters with gorillas were felt to be grossly exaggerated, and his flamboyant manner irritated a number of people. The book which he had published about his travels was also subject to criticism. The frontispiece, a picture of one of his alleged new species of chimpanzee, was said to be a copy of an illustration of a gorilla in a museum in Paris. One of his severest critics was John Edward Gray of the British Museum. His accusations naturally upset Du Chaillu, and in May, 1861, he replied in the columns of the London *Times:*

> Would it not have been more fair of Mr. Gray (of the British Museum), before giving vent to insinuations that I had never visited the countries described, nor collected in those countries my natural-history specimens, to have applied to my friends at Corsica and on the Gaboon, whose names are mentioned in my book? Mr. Gray pretends to be in communication with the missionaries and traders in these parts, and therefore this course would have been more obvious, and he would have saved himself from the imputation of mere calumnies. . . .

Gray replied in the columns of the same newspaper:

"If Mr. du Chaillu had published his works as 'the adventures of a gorilla slayer' I should have taken no notice of it, for the readers of such works like them seasoned to their palate. It is only as the work of a professedly scientific traveller and naturalist that I ventured my observations."

Du Chaillu continued to have problems, many of which were due to his own temperament. On one occasion when he attended a meeting of the British Anthropological Society, he lost patience with the members, and climbing over the benches, he shook his fist in the face of the distinguished and astonished chairman.

Later a young man about town, Winwood Reade, also critical of Du Chaillu's claims, went out to West Africa to check on the explorer's story. Reade, unfortunately, saw only one gorilla, which disappeared into the jungle growth.

Although he had but one brief "contact" with a gorilla, Reade conducted many interviews with the Africans and, on the basis of these, decided that a number of Du Chaillu's claims were false. "M. du Chaillu has written much of the gorilla which is true, but which is not new, and a little which is new, but which is very far from being true."

With all the criticism, it appears that Du Chaillu's descriptions of the gorilla must have been the result of personal observations, and we must give him credit for his taxonomy. Some of his stuffed specimens and skeletons were excellent, and a few are still in good condition in the British Museum and in the Museum of the Royal College of Surgeons in England.

In spite of his limited firsthand experience with gorillas, Winwood Reade came closer to discovering the true nature of the gorilla than did Du Chaillu. Du Chaillu's description of the gorilla as a malevolent jungle monster may have been motivated by his need to establish his courage and hunting prowess and also by his own sense of guilt. The hunter who kills one of these huge, noble creatures could be somewhat disturbed when he observes that the corpse of the animal he has just shot bears a disquieting resemblance to man. To Du Chaillu and the hunters that followed, the gorilla represented a fearsome aggressor who had to be justifiably shot in self-defense.

Du Chaillu's descriptions of his encounters did, however, give birth to the conception of the gorilla as a horrible, fierce, and violent animal, a creature that was oversexed and lusted after human women. In fact, out of Du Chaillu's slanderous portrait of the gorilla was created the fabric of a dozen gorilla myths. *Cassell's Natural History,* published in 1877, obviously influenced by Du Chaillu, described the gorilla as follows:

Travellers in those tropical regions, which are fatal to Europeans, have from the earliest times told of the man-like creatures they had heard of and sometimes seen; and they have associated them in the equatorial part of the continent with human dwarfs, pygmies and monsters. For centuries these degraded human races have been sought after, and now whilst it is admitted that dwarfed men exist, it has come to light that most of the stories which led to the belief in their hideous associates were derived from the existence of large, man-like apes—creatures of dread to the natives—whose traditions are full of credulous anecdotes about them. Hidden in the recesses of the vast forests, where the silence of nature is intense, and moving with great activity, where none can hardly follow, these animals acquired most doubtful reputations, and their ugly personal appearance, so suggestive of violence, was magnified in every way in the eyes of the timid natives.

So dreaded were these apes, and so environed were they with a superstitious mystery, that Europeans had travelled and traded close to their haunts for centuries before one of them was seen by any other eyes than those of the timid negroes. . . .

Sometimes the natives assert, when a company of villagers are moving rapidly through the shades of the forest, they become aware of the presence of the formidable ape by the sudden disappearance of one of their companions, who is hoisted up into a tree, uttering, perhaps, only a short choking sob. In a few minutes he falls to the ground, a strangled corpse; for the animal, watching his opportunity, had let down his huge hind hand and seized the passing negro by the neck with a vice-like grip, and had drawn him up into the branches, dropping him when life and struggling had ceased.

The hideous aspect of the animal, with his green eyes flashing with rage, is heightened by the skin over the orbits and eyebrows being drawn rapidly backwards and forwards, with the hair erected, producing a horrible and fiendish scowl.

The strength of the gorilla is such as to make him a match for a lion, whose strength his own nearly rivals. Over the leopard, invading the lower branches of his dwelling place, he will gain an easier victory; the huge canine teeth, with which only the male gorilla is furnished, doubtless have been given to him for defending his mate and offspring.

This conception of the gorilla as a fierce, hideous monster persisted for more than a century and in fact still exists today. The increased knowledge of the gorilla and the observed facts of its behavior in the wild have had little effect on the original myth of the gorilla as a savage and grotesque hairy ogre. Few people, the twentieth century has pointed out, want the truth when it is dull; what is preferred is an exciting although less accurate alternative. Thus it was understandable and inevitable that the gorilla would be fair game for the arts. No studio artist could dream up a more superlative choice as a movie monster.

Boris Karloff in *The Ape.*

2

The Gorilla as a Movie Monster

By the time motion pictures were introduced at the end of the nineteenth century the public's familiarity with the gorilla myth was well developed. Early filmmakers, catering to the public's fascination with horror, had only to exploit the gorilla's unique qualities for terrorizing an audience.

It was inevitable that the cinema would discover the gorilla. The early success of films dealing with fantasy and the grotesque, especially those of France's film pioneer and ex-magician, Georges Mélies, encouraged producers to scour the world's folk tales and literary classics to come up with enough monsters and unearthly creatures to satisfy the increasing appetite of the new medium. Had the movies not found the already existing gorilla, it is entirely conceivable, as others have speculated, that some sort of similar creature might have been created, for the gorilla was the perfect image of the traditional manlike beast. Huge, hairy, and (by reputation) sexually aggressive, the gorilla was unmistakably designed for fright. So out of the universally accepted gorilla myth developed what was to become one of the cinema's most durable monsters, the

19

stereotyped movie ape: a huge, misshapen brute, bent on destroying humanity and abducting women; a cinematic image of terror that came to represent the public concept of the gorilla on the screen and off the screen as well.

The earliest and most primitive horror films offered at least one and often several gorillas in the cast. The gorillas, of course, were fake. Some were stuffed specimens, but more often they were actors wearing ill-fitting gorilla costumes. The early horror film could easily have kept gorilla-suit makers busy, for it soon became commonplace, in the early silents, to witness the hero chased across the screen by a club-swinging ape, or an actress clad only in a flimsy nightgown being snatched up and carried out of her bedroom by a big hairy monster for the sole and evil purpose of rape. These scenes later became standard for gorilla monster films, particularly the latter, for no self-respecting producer of horror films would ever omit this bedroom confrontation of Beauty and the Beast.

The shadow of Darwin falling across the screen inspired a number of horror films dealing with the theory of evolution. One of the earliest motion pictures to suggest this theme was released in 1908. Entitled *The Doctor's Experiment, or Reversing Darwin's Theory,* its plot is not unfamiliar to present-day horror-movie fans. A mad scientist experimenting on humans changes them into gorillas. He decides to reverse the procedure, but the formula doesn't work. Undismayed by his scientific failure, he is enterprising enough to put his gorilla people to work in a traveling sideshow.

A 1913 vintage movie which employed the ape-into-human theme was based on Gaston Leroux's popular literary work of the period, *Balaoo.* In this early thriller a monstrous gorilla is gradually transformed into a half human by a scientist, Dr. Couriolas. Enemies of the scientist capture the hybrid ape and order it to abduct the scientist's daughter. The human half of this semisimian decides to resist the order, however, but pays for this humane effort with his life.

The monster movies of the 1930's continued their fascination with the gorilla and its human similarities, coming up with theories that would have staggered Darwin and suggesting grisly medical experiments that would turn the stomach of a witch doctor.

In a remake of Gaston Leroux's *Balaoo,* entitled *The Wizard,* a scientist grafts the face of a human fiend onto the features of a gorilla. He trains the disfigured ape to steal and kill, and while the man-faced ape plunders away, the mad scientist chuckles approvingly in his secret laboratory.

Other man-faced gorilla films followed *The Wizard,* including a third version of *Balaoo,* called *Dr. Renault's Secret.* Dr. Renault's secret, obviously, was the dreadful thing he did to the unsuspecting gorilla's face in his basement laboratory.

The first movie gorilla to get a human brain, that of an executed criminal, was ex-wrestler and actor Bull Montana in the 1920 film *Go and Get It.* A rash of films dealing with brain transplants from criminals into gorillas followed, all suggesting that the gorilla was a natural host for a criminal's mind.

The interchange of glands and organs among apes and humans kept many a plot going during this period. Even the great character actor Lon Chaney, known as "the man of a thousand faces," made a film in this genre called *A Blind Bargain.* Chaney, in this tale of horror, enacts a dual role of gorilla and scientist. His disfigurement, causing him to become an apelike monster, is the result of gorilla glands having been introduced into his human body. In one of the few Lon Chaney movies with happy endings, the ape monster takes his revenge on the scientist, by freeing all the other monsters from their cages.

To actor Bela Lugosi, whom we have met on the screen as the bloodthirsty Count Dracula, goes the dubious credit of having metamorphosed more apes into humans and humans into apes than any of moviedom's mad scientists. Some of his transformations, which show up occasionally on television's monster festivals or the *Late Late Show,* are evident in such titles as *Return of the Ape Man,* in which he defrosts a survivor of the Ice Age and hastily changes him into a gorilla. In *Bela Lugosi Meets the Gorilla Man,* we find Lugosi as the mad scientist on a Polynesian island, changing people into gorillas as swiftly as he can capture them. And in the movie version of Edgar Allan Poe's *Murders in the Rue Morgue* Lugosi plays Dr. Mirakle, exhibiting Erik, the gorilla, as a living example of the Darwin theory. Interested ladies patronizing his salon are given belled bracelets, and Erik is loosed at night to trace the tinkling sounds and bring back his victims as experiments for Mirakle's ape-human blood transfusions.

The sex theme, gorillas and women, was played up in the film *Lorraine of the Lions,* which was produced in 1925. This is one of the earlier movies to deal with the fatal and irresistible passion that a beast might feel toward a beauteous woman. The story tells of a ship going to the United States which is wrecked, and a little girl called Lorraine is the only survivor. She is tossed up by the waves on the beach of a deserted island. As in the Tarzan story, she grows up with a group of beasts that includes a pride of lions and also a gorilla whose name is Bimi. Bimi becomes her companion. Many years after, she is rescued, but insists on taking her gorilla with her. He is kept in a cage and when he does eventually get loose, he kidnaps Lorraine. Everyone is horrified, and a great chase ensues. Eventually the poor animal is killed and his inamorata is rescued.

The Beauty and the Beast theme was luridly exploited in a film called *Ingagi,* which was released in 1929 and played to record audiences. Heralded as a true documentary, it played up the gorilla myth, going to the extremes of sensationalism by showing gorillas coming out of the forest to carry off the African women offered them in a tribal sacrifice. Although the big hairy creatures were obviously actors in gorilla suits, the film went to great lengths to represent itself as factual, leading many people in the audience to believe these incredible scenes to be real. Sometime after its release the Federal Trade Commission issued a cease-and-desist order against the producers of the film, Congo Pictures, claiming that *Ingagi* challenged the credulity of science and charging the filmmakers with "nature faking." The Africans in the picture, the Federal Trade Commission pointed out, were not natives of Africa at all, but members of the black community of Los Angeles. Many of the scenes which the producers claimed were shot in Africa were, in fact, filmed in the Los Angeles Zoo. Even the film's title, *Ingagi,* was falsified, the complaint pointed out. Represented as meaning "gorilla" in one of the African languages, a search failed to show such a word used by any of the African tribes. As a result of the action by the Federal Trade Commission, the film *Ingagi* was taken off the screen.

When movies began to talk in the early 1930's, they introduced a new meaning to the word "gorilla," which found its way, finally, to the pages of *Webster's Dictionary.* In the lexicon of the gangster film, the word "gorilla" was used to suggest a big and loutish character, generally a hoodlum or a thug. As a result, the word "gorilla" crept into common use as a slang expression, suggesting any kind of large human male with no brains or morals, but lots of brawn.

One of the most successful monster movies ever made was RKO's 1933 epic of a colossal-sized gorilla, *King Kong.* Its popularity, which has hardly decreased over the years, has been as-

The Phantom of the Rue Morgue.
Warner Brothers, 1954. The late
Charlie Gemora applying make-
up.

tounding. The words "King Kong" or "Kong" have become a
universal household expression, a figure of speech suggesting
anything oversized or powerful beyond belief. The image of
Kong hanging on the Empire State Building in New York has
been reproduced, with varying degrees of phallic suggestive-
ness, in countless numbers of cartoons and advertisements.
One cartoon in a recent travel magazine depicted a huge gorilla
lost on the streets of Manhattan, inquiring of a policeman
where he could find the Empire State Building. The movie itself has developed a "cult" following, taking its place among
such other classics of screen horror as *Frankenstein* and *Dracula,*
with fan clubs and followers that number in the thousands, and
attracting large audiences when reshown in theaters or on television. Some movie critics, pointing to its technical excellence
and to the "integrity" of its central character, Kong, call it the
greatest monster film of all times. To quote one critic, "Kong
was a creature of nobility . . . and, above all, pathos . . . he
was, in a sense, a prehistoric Lear."

Although no filmmaker can claim credit for its final screen
form, the idea for *King Kong* was originally the brainchild of
Merian C. Cooper, a documentary film producer whose combat
flying in World War I had brought him sixty medals and who
was responsible, with an associate, Ernest B. Schoedsack, an ace
Red Cross combat cameraman, for the film. Schoedsack and

Cooper had first met during the siege of Kiev in the Russo-Polish War and had both visited Persia in 1925 to make a film called *Grass,* about the migration of the primitive Bakhtiari tribesmen and their sheep. Later they voyaged to Thailand and made a documentary film called *Chang,* about Laotian tribesmen, including shots of man-eating tigers and elephants stampeding through a village. These two notable true-life classics, *Grass* and *Chang,* brought them worldwide acclaim.

While Cooper and Schoedsack were in Africa filming background footage for a theatrical feature, Cooper became interested in the habits of the gorilla. This inspired the idea of a huge gorilla, of superior intelligence, escaping from captivity and finally running amok in the streets of a large city. He thought also of adding some scenes of the enormous gorilla fighting with prehistoric reptiles which were his neighbors in some remote and forgotten land. For these scenes Cooper envisioned using live gorillas and lizards enlarged by trick photography.

In 1931 Cooper was brought to RKO Studios by its new production head, David O. Selznick. Cooper soon convinced his new boss that the film of *King Kong* ought to be made. Selznick, in agreeing, suggested that Cooper work with Willis O'Brien, the special-effects genius who had done an outstanding job with his unique animation of prehistoric animal models in the motion picture *The Lost World.*

In New York, however, the executives of the company were more hesitant, and in the end Cooper and O'Brien were authorized to produce only one reel. This was to be considered at a sales meeting at a later date. The reel used some of the techniques and models which O'Brien created for *The Lost World* and incorporated a shot showing Fay Wray and a tyrannosaurus. It also utilized for the first time in RKO Studios the back projection technique. When the reel was finally shown at the sales meeting, it turned out to be a sensation. The other sequences which were included in the reel showed sailors trying to cross a ravine on a tree trunk which had fallen across it and Kong shaking the trunk so that the sailors all fell off into the ravine. Another sequence showed a fight between Kong and a tyrannosaurus.

The test was successful in convincing the stockholders to allocate a budget of over $600,000 for *King Kong's* production, an impressive sum for those days. Cooper, with O'Brien handling the special effects, was entrusted with the job of producer and brought in his former associate, Schoedsack, to codirect the picture with him. In his casting for the leads, Cooper contacted actress Fay Wray to play the role of the beautiful blond heroine with whom Kong falls in love. Cooper, it is said, promised the actress that if she accepted the role, she would have as her leading man "the tallest, darkest man in Hollywood."

For a time the famous British writer of mystery stories Edgar Wallace collaborated on the script of *King Kong.* He had come to Hollywood in November, 1931, on a very lucrative three-month scriptwriting contract. He collaborated with Cooper in the early stages of the treatment of the script, but this collaboration was cut short by Wallace's death from pneumonia on February 10, 1932.

The realistic and believable behavior of King Kong, the prehistoric monsters and the tiny, doll-sized replicas of the human actors was accomplished by a technique of animation called stop-motion photography. The lifelike models of Kong and the

Poster advertising *King Kong.*

King Kong, starring Fay Wray, Robert Armstrong, and Bruce Cabot, an RKO Radio picture. Kong battles a pterodactyl while holding Fay Wray in one hand.

King Kong. Kong on exhibition in New York.

Close-up of King Kong model. RKO Studios, 1933.

prehistoric beasts averaged sixteen inches in height and were molded out of sponge rubber, and in Kong's case the model was covered by a shaggy fur. The process of stop-motion photography was a slow and painstaking one. It took an entire day of shooting to get a half minute of the desired action on film. First the model was set in place and one frame of film exposed. Then the model's position was changed, a hand thrust forward slightly or a foot advanced almost imperceptibly. To get just one of Kong's footsteps photographed on the film took twelve different movements and exposures. A single sequence, the fight between Kong and the pterodactyl, took seven weeks to photograph. When the animal shots were completed, the human actors took their places in front of a screen and the animal scenes were projected onto the screen from the rear. The actors were then photographed reacting to and synchronizing their movements with the action of the monsters on the screen. When shown later in theaters, this effect of combining the live action of the actors with the prefilmed movements of the monsters provided a remarkable illusion of reality.

For close-up shots a full-sized torso and hand of the giant Kong were built. The huge ape's eight-foot-long hand and arm were operated by a steel rod and a cable attachment within the arm, providing movement for his hand and fingers, and the entire arm could be raised or lowered like a crane.

One of the many problems that arose during the making of the picture was trying to get the proper vocal sounds for a fifty-foot ape. The amplification of an actual gorilla's bark failed to suggest anything coming from a creature the size of Kong. Recording the gorilla's bark backward, however, and at a slow rate of speed, was a distinct improvement, but the final solution came as a result of playing the reverse sounds of the gorilla through a specially constructed, 25-foot-square sound box. No one can argue on seeing the movie that Kong's anguished cries aren't coming from a prehistoric fifty-foot ape.

The story of *King Kong* retells the age-old legend of Beauty and the Beast. The film opens, in fact, with a quotation from an old Arabian proverb: "And the Beast looked upon the face of Beauty, and lo! his hands were stayed from killing. And from that day forward, he was as one dead." Carl Denham, an explorer and producer of motion pictures, learns of an island off the coast of Sumatra where prehistoric animals are still living. He takes with him, as his leading lady, the beauteous blond actress Ann Darrow, and when they arrive at their destination, the hitherto unexplored Skull Island, she is soon captured by the Sumatrans and presented to King Kong, the tremendous gorilla, lord and ruler of the island, as his bride. Kong claims his beautiful prize, but instead of destroying her, he falls in love with her. As he returns to his primeval forest with the helpless Ann, he is forced to defend her from two prehistoric beasts, a tyrannosaurus rex and a pterodactyl. Finally the ship's crew are able to find and rescue Ann, and Kong is captured with the aid of gas bombs. Kong is brought to New York by Denham, who plans to exhibit him as "the eighth wonder of the world." As news photographers are taking pictures of Ann Darrow, Kong believes the flashbulbs are meant to harm her; he breaks out of his chains and blunders through the streets, wreaking havoc as he goes. Discovering Ann in her hotel room, the giant Kong picks her up, bed and all, and clutching her tightly, he swings from building to building, stopping momentarily to do battle with an elevated train. Then, reclaiming his prize, he begins his journey of destruction through Manhattan, headed for the safety of the city's tallest skyscraper, the Empire State Building. He makes it to the very top of the building and tenderly deposits Ann on the highest ledge. Suddenly he hears the roar of airplane motors, and a squadron of air force fighters dive on the giant ape, strafing him with machine-gun bullets. Kong is defiant, but wave after wave of planes swoop down at him, their machine guns blazing. Soon Kong is mortally wounded; bleeding from chest to neck, he is unable to hold onto the spire of the building. He picks up Ann for one last sad look and painfully, but tenderly, returns her to her perch. For the great shaggy King Kong, lord and master of Skull Island, it is all over. Clutching at his throat, he totters and falls to his death below. "Well, Denham, the airplanes got him" remarks a passing policeman. But the producer shakes his head in disagreement. "It wasn't the airplanes. It was beauty killed the beast."

The smashing success of *King Kong* called for a sequel, and Cooper and Schoedsack came up with *Son of Kong*, which they rushed into theaters for the Christmas season of 1933. The film made no attempt to explain how Kong, supposedly millions of years old and unmated, could have an offspring, but junior turned out to be such a clownlike character that it was released as a "serio-comic phantasy." In the film, Denham, plagued by debts for damages done to New York City by Kong senior, returns to Skull Island, hoping to find Kong's secret treasure. There he finds King Kong's offspring, a huge albino gorilla, who becomes attached to the party, following them around like a faithful puppy, but showing signs of being "a chip off the old block" when prehistoric monsters threaten. At the film's end an earthquake and gigantic tidal wave cause Skull Island to sink, and the self-sacrificing young gorilla holds Denham and his party in his giant paw until a rescue boat can reach them—and then he drowns.

Another film whose tenderhearted giant gorilla theme in some ways paralleled *King Kong* was Merian C. Cooper and Willis O'Brien's *Mighty Joe Young*. Released in 1949, sixteen years after *King Kong*, and employing *Kong's* same astounding

Mighty Joe Young and his owner. RKO Studios, 1949.

special-effects technique, *Mighty Joe Young* won for O'Brien a richly deserved Academy Award. Its story has been described as being more of a children's fairy story than one of monsters. Joe Young is a huge black gorilla who has been raised from infancy by a young and beautiful girl, daughter of an African ranch owner, who keeps the massive ape around as a household pet. A showman, traveling with an American rodeo company in Africa, is attracted to the girl and persuades her that both she and her pet gorilla belong in a New York nightclub. On the opening night of their engagement Joe Young appears onstage holding the girl and the piano aloft while she plays the ape's favorite song, "Beautiful Dreamer." Their nightclub debut is suddenly cut short when the noise and the lights and the gibes of a drunken patron finally cause Joe Young to go berserk, and he wrecks the nightclub with its African motif and frees the caged lions which have been provided for atmosphere. Joe's life is spared, however, for as he is about to be mowed down by the police, he manages to save some children from a burning orphanage. Satisfied that show business is not for him, Joe returns to the peace and quiet of his former life in Africa.

King Kong, Son of Kong, and *Mighty Joe Young* were notable exceptions to the stereotyped movie ape and the gorilla myth. While these films must be regarded as fantasies suggesting a distortion of the size and character of the true gorilla, they generated an audience sympathy for the unhappy, lovelorn Kong, an affection for the cuddly albino son of Kong, and a genuine concern for Joe Young's safety and welfare. In spite of

Shot of model of Mighty Joe Young. RKO Studios, 1949. (Courtesy of Robert H. Burns.)

these exceptions to the typical gorilla-monster genre, each a plea to consider the gorilla's gentle nature, the film industry continued to grind out the age-old gorilla myth, offering the public such titles as *Bride of the Gorilla, The Beast That Killed Women, Gorilla at Large,* and *The Bride and the Beast.*

The film *Mogambo* had some terrifying and sensational gorilla sequences in it which did not depend on men in gorilla suits. They were real. The man who was responsible for these shots was a cowboy called Yakima Canutt. Canutt had been a rodeo performer, a Western film star, and a stunt man, and Hedda Hopper wrote about his efforts to get these gorilla shots in the Los Angeles *Times* way back in 1953. MGM, which made the film, did not really believe that Canutt could get the shots it wanted, but he did, and it cost $10,000 a minute of film time, making a total of $150,000 for a fifteen-minute sequence. Canutt was not that knowledgeable about gorillas, and he had no idea whether it was possible to call the bluff of a charging gorilla. He was heartened, however, when he saw an encounter between an African and a gorilla which resulted in the latter yielding to the African and returning to the jungle. Later he talked to another African, who claimed that his hand had been bitten off by a gorilla. The African said that he found a gorilla tearing up one of his banana trees to get the pith out of it, and he told the gorilla, "It's my tree and if you don't go away, I'll throw a spear at you." The gorilla ignored this remark, and the man threw his spear at it. As it pulled the spear out and bit the hand off the African, the gorilla said, "I'll never come back to see you again. You're a bad man."

The Africans speak of the gorillas as if they are men and claim, as the above story shows, that they can speak. The men claim that they are at war with the gorillas.

Having eventually won the cooperation of the Africans, Canutt started his gorilla-filming operation. The first problem was to get the supplies into the gorilla country. They had to be hauled in by a truck from a town 700 miles away, and they included over a mile of wire netting and hundreds of iron posts. Canutt enlisted the help of 40 trackers and 600 hunters, each of which he paid fifty cents a day. He then found he was expected to pay their wives the same amount, which he did in order to keep the peace, but he never understood why he was supposed to do this.

To be sure, he had wild gorillas to film when he wanted

Konga. Herman Cohen Productions, 1961.

them. He had his trackers locate a group of thirteen gorillas and then build a wire fence a mile and a half in circumference around them. At one end of the enclosure the jungle was hacked away and a clearing produced, which was the actual site of the filming. Then a series of fences was thrown up across the enclosed jungle, slowly squeezing the gorillas closer and closer to the cleared area. At this point the Africans, knowing the job could not now be completed without them, went on strike for double pay. Canutt had to give in and raise their wages to a dollar a day.

When the shooting began, the cameraman was enclosed in a wire cage, but he had the jitters all through the filming. Near the camera a man-shaped dummy was hung, which was shaken from time to time to give the gorillas an opportunity to charge.

Crash Corrigan in *DOA*. Paramount Pictures.

Eight gorilla charges were filmed, the big male getting within ten feet of the camera at one time.

While all this was going on, Ava Gardner and Clark Gable were making the rest of the film at a site 1,500 miles away from the gorillas. The gorilla film was eventually flown to London and used as background for the actors.

Another remarkable and very recent film showing gorilla charges was made by John Hemingway in association with Adrian Deschryver near Bukavu in Zaire and was shown on network television in the United States. In one of these charges a male gorilla hit Deschryver on the back with his hand. If it had been a full-fledged gorilla whack, it would have flattened the recipient on the ground, but apparently the blow barely brushed Deschryver's jacket. Since he was well known to this particular group of animals, it is possible that the gorilla did not really want to hurt him. In another sequence in the film Deschryver, holding a tame baby gorilla in his arms, strolls into an area where the gorillas are feeding. The sight caused great excitement among the wild gorillas, and the number one male charged Deschryver, who, thinking he was going to be attacked, dropped the baby gorilla and stepped back. The gorilla grabbed the baby, pulled it away, then left it and ran back with

Darkest Africa. Republic Studios.

the others. Then it made another charge, this time picking up the baby and returning with it to the rest of the group. They gathered around the youngster, examining and grooming it. But the baby gorilla was too young to go back in the wild. There was no lactating female to cuddle and nurse it, and within ten days it was dead.

The movie ape found its image somewhat reversed in the late sixties when Twentieth Century-Fox Studios released a series of films based on Pierre Boulle's novel, *Planet of the Apes.* This motion picture series, which had great success at the box office and was until recently a weekly television series on the Columbia Broadcasting System, deals with a simian society in the year 3955, where gorillas, orangutans, and chimpanzees share a role in the advanced social order.

A spaceship, spinning through galactic infinity, crash-lands on an unfamiliar planet in some apparently unknown zone beyond the solar system. Launched from Cape Kennedy some twenty centuries earlier, the craft has traveled at the speed of light, and since its astronauts have sealed themselves in a kind of suspended animation in airtight bunks, they find they have aged only eighteen months.

When rounded up and captured by a mounted patrol of gorilla policemen, the astronauts realize they have landed in a strange simian world where the evolutionary processes have been working counterclockwise. The ape is civilized, clothed, cultured, and capable of speech. Man is a wild, unintelligent, naked brute. It is the ape, in the topsy-turvy society, that dominates man. He cages man, dissects his body for experimental research, and mounts him for display in museums. Human population is restricted in numbers, for the apes are concerned lest man overrun and destroy ape society. In the ape hierarchy the gorillas are the enforcers of the law, the police and the military; orangutans serve as ministers of state, judges, and administrators of the law; chimpanzees are the intellectuals and professionals, sometimes suspect because they question their society's doctrines.

The *Planet* films were full of biting irony—a science-fiction fantasy punctuated with penetrating commentary on human values. Its satire touched familiar themes. A gorilla is eulogized at a funeral ceremony as having said, "I never met an ape I didn't like." A chimpanzee repeats the words "Human see, human do"; and a tribunal of orangutan judges strikes the classic simian pose resembling "See no evil, hear no evil, speak no evil."

The story behind *Planet of the Apes* is one of a prodigious, uphill battle for its producer, the late Arthur P. Jacobs, a former public relations man who once represented Marilyn Monroe.

Jacobs obtained the screen rights to *Planet* while it was still in manuscript form, but every movie studio he contacted in the United States and in Europe refused to consider it. They felt that putting a bunch of talking apes into a serious drama just wouldn't work. Jacobs refused to give up. He had noted science-fiction writer Rod Serling write a screenplay and persuaded actors Charlton Heston and Edward G. Robinson to make a brief test while appearing in uniquely designed simian makeup. Twentieth Century-Fox Studios, impressed with the film test, reconsidered. They finally agreed to make the film.

From the beginning the film's technicians faced monumental problems. They had to deal with a cast in which nearly all the leading characters appear as apes. These actors were not to masquerade in gorilla suits or other ape costumes, as was customary in other ape-theme films; they would have to come through as living, breathing gorillas, chimpanzees, and orangutans. This called for the design of a lifelike facial mask, which, despite its elaborate disguise, would reflect the subtlest emotional reaction on a player's face, allowing him to move facial muscles and change expressions at will. The standards set for these simian facial transformations went beyond anything previously established for a massive cast effort; the actors were not only to look like apes, they must be accepted by the audience as intelligent beings, capable of speech, thought, and reasoning, even artistic and scientific achievement.

Possibly the largest makeup force in movie history, nearly eighty makeup technicians, worked under Fox's chief makeup specialist, John Chambers, a former surgical technician who had helped repair the faces of soldiers wounded during World War II.

But makeup artists weren't all the technicians needed to accomplish the required ape transformation. Chemists, sculptors, and wigmakers were called in. The chemists, experimenting with new rubber compounds, developed materials that permitted full facial mobility and at the same time allowed the actor's skin to breathe inside the heavy outer layer of ape makeup.

During the early stages the makeup required six to seven hours to apply. It took three hours to remove, but no actor, the studio realized, could survive an eighteen-hour day, five days a week, so new techniques had to be devised to speed up the application and removal of the ape disguises. Eventually, a sizable army of technicians was trained to apply the makeup in three to four hours. It took one to two hours to remove it.

The makeup is not really a mask; it has been described as an "appliance." Dan Streibeke, who was cocreator of these appliances, explains that "the only objects we can save to use another time are the ears. As for the rest of it, my makeup artists started from scratch at 5:00 A.M." Streibeke calls his men makeup artists because he regards them rather like sculptors, and the actor's face is used in the process of making up the first step of a life mask. For this purpose they use a material called Jelltrate, which is used by dentists to make impressions for false teeth and for bridges. An impression is then made of the face from dental stone. Then clay is used to add the features of whatever type of ape is to be portrayed by the actor—either the gorilla, the chimpanzee, or the orangutan. The dental-stone impression is fitted together with the Jelltrate impression, and foam rubber produced by a special formula is pumped in between the two. Once it is set, the rubber is taken out and is then put in an oven for about six hours at 200° F. This is known as the basic appliance. It is then sprayed with coloring material which is attracted to rubber. It sticks very well to it, and all the apertures, such as those for the eyes, are cleared of bits and pieces. Then a channel is cut through the roof of the mouth. In this way the actor is able to breathe. One of the advantages in using this material and this type of appliance is that it is mobile, and the actors can use it to convey emotion. Roddy McDowall, who is the star of the show, says, "One has to learn to exaggerate the facial movements in order to energize the appliances. It is rather like the old trick of patting the stomach with one hand and circling the head with the other, but it can be done. Otherwise, all one would have is an expressionless, lifeless mask."

Each morning the massive cast of ape actors spent hours in makeup chairs as thin-layered masks, plastic noses, anthropoid jaws, and hair were applied. They were fitted with an extra set of teeth set in their jaws that protruded beyond their own. Eating was a problem; liquids could be sucked through a straw, but many of the actors learned to use mirrors to guide their forks past a false oral cavity into their own mouths. Somehow the actors got used to living with the discomfort of an ape face. "After a while you forget you look like an ape," said actor Maurice Evans, who played Dr. Zaius, the skeptical orangutan.

Out on location on a hot day actors wearing these appliances and costumes have often lost as much as ten pounds in weight. Roddy McDowall, who in the television series takes the part of Galen, the chimpanzee sympathetic to the astronauts, had on occasion to keep his appliance on for thirteen hours. It cannot be removed while the actors are working.

For masterminding and supervising the prodigious makeup efforts for *Planet of the Apes* chief makeup specialist, John Chambers, won a special Oscar, one of the two awards for original makeup in Academy history.

Bob Burns is an actor who has made a career of playing the

Tarzan and His Mate. Metro-Goldwyn-Mayer, 1934. Gorillas and chimpanzees figure in this shot.

Tim Tyler's Luck. Monogram Studios, 1937.

The ape family of Janos Prohaska and son.

The late Janos Prohaska, famed movie gorilla, who also played Cookie the Bear on television, was killed in a tragic plane crash while returning from a film location for TV's *Primal Man* series. His pose here with the twins was meant to represent his desire to portray the gorilla as a gentle animal.

Lucille Ball and friend. From *The Lucy Show*, a CBS television program.

Bob Hope, Bing Crosby, and gorilla friend in *Road to Bali*. Paramount Pictures, 1952.

gorilla. He was recently interviewed and talked about the trials and tribulations and rewards of being a gorilla. He developed a great affection for his gorilla suit and personalized it, always referring to it as "him." Once inside the suit, Bob found it difficult to communicate in "humanese" any longer and lapsed into snorts and grunts. He does not feel rough and aggressive in the suit, but warm and lovable. Of his feelings inside the suit, he said, "It's just like I become another person almost, another personality or something, because I feel for the gorilla so much, I love the character so much. I believe I would play the part twenty-four hours a day if I could. I just wanted to be a gorilla man." Bob had studied gorillas and their movements and wanted to play the gorilla in a movie in which he could actually move like a gorilla, but he has always run up against directors who have their own fixed ideas as to how gorillas should move, and it was not the way real gorillas did. So the gorilla man had to lumber around upright on his two legs in a semicomic fashion; according to him, directors and producers just don't seem to know the difference. Janos Prohaska, who was killed with a number of colleagues in a tragic plane crash and who also played a gorilla, had reached the stage where he refused to burlesque a gorilla anymore, and he had arguments with directors about the way real gorillas would act.

Bob Burns makes his own gorilla suit. The fabric is made of nylon and costs fifty dollars a yard. It is now difficult or impossible to get this material, so the gorilla suit is very precious since it cannot be repaired. The inside of the head is an exact negative of Burns' face and fits over it like a glove, and when he gets his makeup on, it is impossible to tell where the makeup stops and where his real eyes start. On one occasion Bob was talking to a young girl about his makeup and she wanted to know how the eyes go in there. Bob said they were his own eyes, but she insisted they must be some kind of mechanical eyes.

One of the problems about Bob Burns' interpretation of the gorilla was that he had blue eyes and gorillas have brown eyes. Only one gorilla with blue eyes is known and that is Snowflake, the white gorilla, but gorillas' eyes do vary in shade. Despite the difference between blue and brown eyes, Bob said that no one had ever noticed his eyes and he assumed that people are too busy taking in all the movements and activities of the human animal to do so.

The Burns gorilla took part in an opening of a swimsuit establishment and wore a bikini and a little girl fell in love with this combination. She was only about seven years old and sat with the gorilla, combing his arm and hugging and kissing him. Wearing a gorilla suit in public is, however, a dangerous activity. There are bullies who will try to kick the gorilla or hit it on the head. Kids in general seem to want to hurt the animal, and they are often egged on by the adults. It is difficult to decide whether the people think the animal is real and they want to hurt it or whether they realize that it is a man in a suit and want to humiliate him. People seem to react a lot stronger to the gorilla than they do to any of the other monsters in costume, but even such an innocent figure as Mickey Mouse in Disneyland has been stabbed six times. One incident which was very disturbing happened to another human gorilla, George Burrows. He was making a picture called *Gorilla at Large* in a Pacific Ocean park, and a number of marines and soldiers had been let into the park for the filming for atmosphere. One of them tried to set Burrows in his gorilla suit on fire. The suit was made of yak hair, and if the attempt had been successful, he would have gone up like a Roman candle and it would have been impossible to get Burrows out of the suit in time.

Some of the other gorilla men who took part in films were Emil van Horn, who was the gorilla in the film *The Ape Man* with Bela Lugosi, and Charlie Gemora in *Phantom of the Rue*

Ramar, a young gorilla performer, one of the animal newcomers to the world of show business.

Morgue and *Africa Screams.* In *Phantom* the long shots were of Charlie in a gorilla suit, and the close-ups were of a big, living chimpanzee. The directors were probably trying to get all the fierce mouth movements of a real animal, showing the teeth, etc. Bob Burns points out that this was back in the thirties, when the audiences were very unsophisticated about apes and probably no one knew the difference. They are a great deal more sophisticated now; even the old myth of the obsession of apes with human females is not as strong as it once was, and Bob Burns asserts that the old film situation in which a man in a gorilla suit carries a girl around and drags her up to the tree-tops would not "go down" anymore with the general public. Part of this, he thinks, is due to the prevalence of so many animal shows on TV, which have given people a much better idea now of what real animals are like.

Filmography of American Gorilla Movies

1908: *The Doctor's Experiment, or*
       Reversing Darwin's Theory, Gaumont
1908: *Sherlock Holmes in the Great Murder*
       Mystery, Eclair
1913: *Balaoo,* Eclair
1914: *The Miser's Reversion,* Thanhouser
1914: *The Fakir's Spell,* studio unknown
1920: *Go and Get It,* First National
1922: *A Blind Bargain,* Goldwyn
1925: *Lorraine of the Lions,* Universal
1926: *Unknown Treasures,* Sterling
1927: *The Wizard,* Fox
1927: *The Gorilla,* First National
1928: *The Leopard Lady,* Pathé
1929: *Stark Mad,* Warner
1929: *Circus Rookies,* MGM
1929: *Where East Is East,* First National
1930: *The Gorilla* (remake), First National
1931: *Murders in the Rue Morgue,* Universal
1932: *Island of Lost Souls,* Paramount
1932: *The Monster Walked,* Universal
1933: *The Thing with Two Hearts,* Twentieth Century-Fox
1933: *King Kong,* RKO
1933: *Son of Kong,* RKO
1933: *So This Is Africa,* Columbia

1935: *Tarzan and His Mate,* MGM
1936: *Darkest Africa,* Republic
1937: *Tim Tyler's Luck,* Universal
1939: *The Gorilla,* Fox
1940: *The Ape,* Monogram
1941: *The Monster and the Girl,* Paramount
1942: *Dr. Renault's Secret,* Fox
1943: *The Ape Man,* Monogram
1943: *Captive Wild Women,* Universal
1944: *Jungle Woman,* Universal
1944: *Nabonga,* PRC
1944: *Return of the Ape Man,* Monogram
1944: *The Monster Maker,* PRC
1945: *The Monster and the Ape,* Columbia
1945: *White Pongo,* PRC
1946: *Spook Busters,* Monogram
1948: *Who Killed Doc Robbin?,* Roach
1949: *Mighty Joe Young,* RKO
1949: *DOA,* United Artists
1949: *Master Minds,* Monogram
1949: *Africa Screams* United Artists
1951: *Bride of the Gorilla,* Broder
1952: *Monster Meets Gorilla,* Broder
1953: *The Killer Ape,* Columbia
1953: *Mogambo,* MGM
1954: *The Phantom of the Rue Morgue,* Warner
1954: *The Bowery Boys Meet the Monsters,* Monogram
1954: *Gorilla at Large,* Fox
1954: *The Bride and the Beast,* Allied Artists
1961: *Konga,* American-International
1963: *The Ape Woman,* Champion
1965: *The Beast That Killed Women,* Mahon
1966: *Ghost in the Invisible Bikini,* American-International
1968: *Planet of the Apes,* Twentieth Century-Fox
1970: *Skullduggery,* Universal
1971: *Escape from the Planet of the Apes,* Twentieth Century-Fox
1972: *Conquest of the Planet of the Apes,* Twentieth Century-Fox

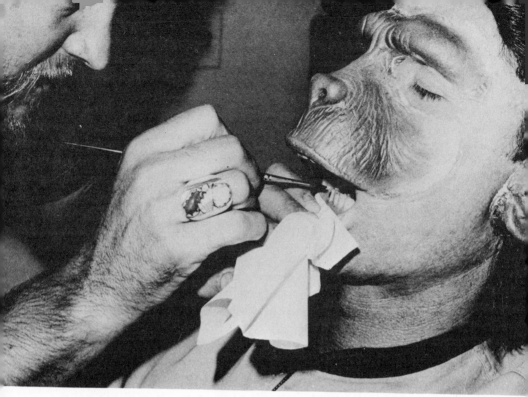

Planet of the Apes. Putting on an ape appliance.

Planet of the Apes. Completing the ape makeup.

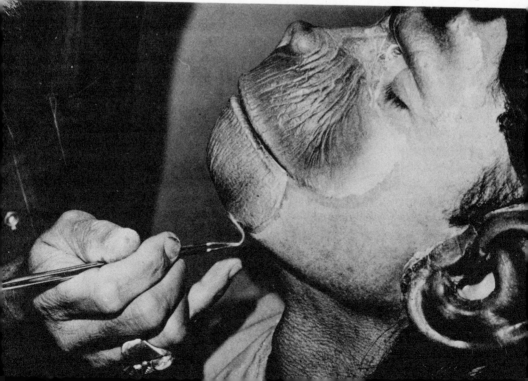

3

"Show Business Is No Business for Gorillas"

WHEREAS chimpanzees have adapted spectacularly to show business, having trouped and starred on the stage and screen and in nightclubs, the circus, ice arenas, and even had their own television series, gorillas seldom see the spotlight. It's not that these big apes don't like to ham it up now and then, but there are a number of reasons why real gorillas are not likely to be found on a theatrical agent's roster. First of all, they cost from ten to twenty times as much as a chimpanzee—probably $10,000 to $15,000—and they are more difficult to train, but most significantly because they are on the endangered list and cannot now be imported to the Western Hemisphere.

One of the very few nightclub acts to feature gorillas was Professor Jon Berosini's mixed ape attraction that played, for several years, at the Circus Circus Casino in Las Vegas. In the professor's simian troupe were three chimpanzees, two orangutans, and two gorillas. There were a number of acrobatic routines on the high wire and trampoline using the chimps and gorillas, and the gorillas proved just as agile as the chimps. In another routine the professor led a full-grown orang to a drum, and while two gorillas wearing an exotic African plumed headdress beat the tom-toms, the orang danced a sort of Hawaiian hula on the drum. The act was climaxed with the chimps spinning on revolving arms and the orang doing fast pinwheel turns while spread-eagled on a large vertical wheel. It was a most impressive act and reflected a prodigious job of training by Professor Berosini.

One of the most extraordinary feats of animal training and one of the most incredible sights of its kind ever witnessed was the show put on by three trained gorillas at the zoological gardens in the city of Nagoya, Japan.

These gorillas, having now all reached maturity and a bit too old for their stage shenanigans, will give only a token performance now and then, but only a few years back they were the delight of school children of Nagoya, performing to the amazement of world travelers and all of Japan.

The gorillas presented two shows a day, and each was directed by Rikizo Asai, the gorilla troupe's trainer and keeper. The performance began with Asai, wearing overalls and a hat with a large floppy brim, leading the three gorillas out onstage, each walking behind the other in a chain and each holding his hand on the shoulder of the preceding gorilla. This, of course, presented a very amusing as well as a rare sight. Trainers in circuses and zoos have made fairly close contact with a single large gorilla, but certainly no one has been on such familiar terms with three large gorillas as Asai. The show continued with the animals putting on turned-down sailor hats and sitting down at a table and enjoying a tea party. Asai would chew food and then transfer it from his mouth into the lips of one of the gorillas, a technique gorilla mothers often use in the wild to give food to their children. It is possible that Asai was able to establish the tremendous trust and close relationship with these big animals due to his capacity to act as a mother figure.

The act went on for a full half hour and included a hilarious weighing-in bit, an exhibition of Japanese sumo wrestling with tiny Asai trying to grapple with his three young giants and they, good-naturedly, letting him, and a musical number featuring the gorillas on percussion instruments, and in one side-splitting number the female gorilla stood out in front of this remarkable rock orchestra and did the twist.

Obviously Asai's trained gorilla troupe caused considerable surprise in the international zoo community. There had always been some doubt that gorillas could be trained to work together in a group situation. Asai proved otherwise. His training methods were unique—they had to be—and he had to have the patience of a dozen Jobs. Asai began working with his gorillas when they were toddlers not even a year old. He immediately tried to gain their trust by acting out various roles, most significantly as parent and also as leader or dominant male. Both resulted in a strong rapport; he gained their confidence and was able to inculcate a stern sense of discipline, which are the basic requisites for training most performers, animal or human.

48

"Gorillas," Asai tells us, "are quick learners, but they are also very nervous, so confidence is extremely important in learning each new routine." Because they are so nervous, gorillas can become frightened very easily with new and strange objects. Therefore, it is very important in the beginning to get them familiar with the object that they are going to use in the show. When, for example, they first saw a drum, they were apprehensive, but became really frightened when Asai thumped on it, and they defecated profusely. To help them get over their fear of the drum, Asai took some bananas, squashed them, and then smeared the bananas over the drum. Then when the drum was shown to them they would smell the banana on it, move in to smell it much closer, and then proceed to lick the drum. Very soon they lost their fear of the drum completely. This was also used as a technique to accustom them to other types of sounds of foreign objects.

The three gorillas used in the show had their own individual characteristics. For instance, not all of them could adapt to playing the drum. Poopi, however, became the master of the drums and was also able to play the trumpet. He mastered these two tricks within a couple of hours. On the other hand, Oki, who was not very good on the drums, was given a harmonica. He picked it up in a very natural fashion and started to play a tune on it, or at least his idea of a tune. Interestingly enough, what he played actually had a certain amount of rhythm to it.

Gorillas are not only big animals; they are enormously strong. Asai was aware that one day they would be too much for him to control. As a trainer, however, he felt that the act, although it was entertaining for zoo visitors, was also important to the mental health of the animals themselves and gave them something to do instead of languishing in boredom in a cage. Asai feels that animals which have been trained and take part in a show are much happier animals and are much more likely to indulge in successful breeding. Another important aspect of the show was, according to Asai, that "it also gave the public an opportunity to observe the intelligence and character of the animals closest to man."

Robert E. Noell, who with his wife runs the Noell's Ark Chimpanzee and Gorilla Farm near Tarpon Springs, Florida, has been working with and raising gorillas for a period of twenty to thirty years. One of his show-business pets is a 600-pound male gorilla called Tommy, who is now fifteen years old. He is an enormous creature, probably weighing the equivalent of three football players. He has been known to put a dent in a car with one backhander, scarcely noticing what he is doing. Most people who come in contact with him are terrified by his size and apparent strength, which is obvious when he uses it. Yet he can be very gentle and even lovable. The Noells got their first gorilla in 1950, but unfortunately she lived for only a year. Later on they bought a male, and he also survived for only a year. The third one, however, lived for four or five years. After only a few years they also lost two more. They paid about $5,000 for each animal. Tommy has been with them for ten years and is one of the few privately owned gorillas in this country.

Sometimes Tommy gets away from his home. On one occasion he tried to direct the traffic coming down the road near his home and created all kinds of confusion. On another day he wandered into the middle of a group of railway workers in the vicinity of Tarpon Springs and created a state of panic among them. According to Dick Bothwell and David Coleman, the journalists who wrote about him, the railroad workers demonstrated in an involuntary fashion the process of seeing how many people could simultaneously cram into the nearest parked car in the shortest possible time.

When Noell and his wife went off on a tour of the country with a traveling show, they always took one of their gorillas with them, particularly Tommy, so that he could indulge in exhibition wrestling matches with volunteers from the audience. Bothwell and Coleman reported that "the animal's huge black hands palm a man's head as a pitcher would palm a softball." Part of the act was to let Tommy take Mr. Noell's arm between his jaws and give him a mock bite. Noell said, "When we're on the road, I go in the cage with Tommy every night. I always tell the audience this may be the last time! But he is gentle. We brought him up like a child. Got him when he was a few months old. Gorillas live to be forty-five or fifty years old. He may be brought up like a child, but at four hundred fifty to six hundred pounds he can be a very dangerous child, particularly when he also stands five and a half feet in height."

Gorillas occasionally have periods of irritation and can do a lot of damage. A backhander from one of the Yerkes Primate Center gorillas, Ozoum, once nearly knocked the tip off a man's index finger. The Noells have done a remarkable job in training Tommy, and he has shown that he has considerable ability in solving problems, even those in which tools have to be used.

50

Using sticks or stacking boxes on top of each other to reach food which is suspended out of his reach does not create a problem even though this type of activity is not regarded as being a gorilla's particular forte.

During training Noell always rewards Tommy. This, however, became rather expensive for the Noells because Robert Noell always used to keep candy in the pocket of his shirt, and Tommy knew where to find it. He very often removed not only the candy from the shirt but the pocket as well. Also, when the two wrestled, the shirt was usually torn to pieces. So Noell made a point of buying his shirts by the dozen from Goodwill, paying in 1969 about fifteen cents each for them. Tommy was a mountain gorilla, and if he got angry or frightened, he was known to rush around hitting anything within his sight. A casual bang on Noell's parked car once resulted in a $50 dent. Noell often used to take Tommy for walks, "leading him on a chain, but mostly it seemed to be Tommy taking Mr. Noell for a walk." It has been said that "Noell looks like a small boy trying to walk a Great Dane," even though he is a pretty husky man himself, weighing 200 pounds.

Robert H. Burns (movie ape) and Art Long in outdoor attraction at Magic Mountain Amusement Park in Southern California. (Courtesy of Robert H. Burns.)

Berosini Jungle Fantasy at Circus Circus Casino, Las Vegas, with gorillas, chimpanzees, and orangutans. (Courtesy of Professor Jan Berosini.)

Berosini Jungle Fantasy at Circus Circus Casino, Las Vegas. Two gorillas and an orangutan put on a rhythm act. (Courtesy of Bobby Berosini.)

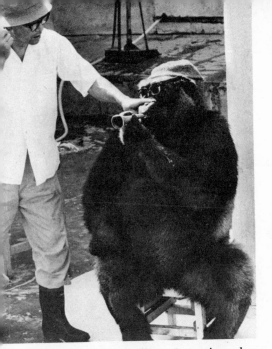

The Nagoya gorillas performing. (Courtesy of Rikizo Asai and the Nagoya Higashiyama Zoo.)

The Nagoya gorillas. Learning the trumpet. (Courtesy of Rikizo Asai and the Nagoya Higashiyama Zoo.)

A reporter once visited the Noells and described his experiences there. He drew a picture of his first view of Tommy, who came stumping along on arms like tree trunks with what appeared to be a scowl on his black face. He was just a mass of sheer muscle, with a beetle brow and underneath it brown eyes that looked at the reporter with a cool, rather indifferent glance. It seemed a surprising thing to find a big gorilla like this running around in somebody's backyard, but this, as the reporter pointed out, wasn't an ordinary backyard. Mr. and Mrs. Robert E. Noell raise chimpanzees, some of which they sell for psychological experiments only; others are raised as performers to box and wrestle with humans when the Noell's Ark show takes to the road.

When he is out of his cage, Tommy has a small chain around his right wrist. His owner and keeper, Bob Noell, holds onto the end of it. It seems to be largely a gesture since there is very little doubt that anyone could stop Tommy if he really wanted to go his own way. On one occasion the animal spotted a carton of milk on the other side of the yard and lumbered over to it with Noell tagging along at the end of the chain as if he were an appendage. Reaching the carton of milk, Tommy sat on the

A Noell gorilla. Mrs. Mae Noell nurses a "baby." (Courtesy of Mrs. Mae Noell.)

A Noell gorilla. Gorilla Tommy regards a friend. (Photo by Norman Zeisloft.)

A Noell gorilla. Gorilla Tommy and friend, Robert Noell. (Photo by Norman Zeisloft.)

bench and drained the milk from it, and Noell sat alongside talking to him. It is difficult to say what was actually going through the great ape's mind at that time. He put out a black hand which was about the size of a dinner plate and put it around Noell's head and neck and then pulled him to him with a great hug. He took Noell's arm, opened his mouth, and put his arm in his mouth between his teeth. Noell called out to him, "Look out now, Tommy. Don't be rough." Fortunately Tommy

A Noell gorilla. Tommy takes a drink. (Photo by Norman Zeisloft.)

Tommy positions Mae Noell in the direction he wants her to go and then bites her on the bottom to start her off. (Photo by Norman Zeisloft.)

A Noell gorilla. Tommy and Robert Noell share a banana. (Photo by Norman Zeisloft.)

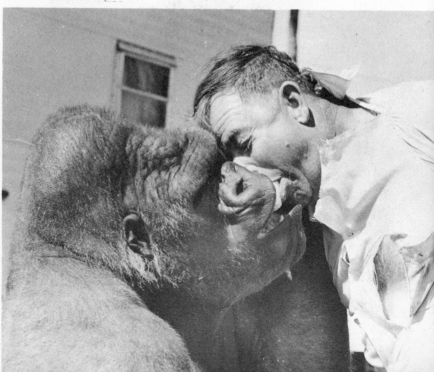

was only doing mock bites on the arm, and eventually Noell got his arm back. Then they engaged in a spell of wrestling. Noell sometimes has to speak sternly to Tommy to remind him that he is a big strong boy. In a wrestling competition Noell's major interest is to stop Tommy from either stepping on him or rolling on him. As he said, "It's bad enough if you have a big man roll on you, but to have a gorilla weighing more than five hundred pounds doing the same thing ceases to be fun." Mrs. Noell said that Tommy once stepped on her head when he weighed only 200 pounds, and she could almost hear the bones crack. The biggest worry is to stop Tommy getting excited because gorillas when keyed up run about in a rather uncontrolled way, striking at anything as they go past. A backhander from a gorilla under these circumstances, particularly one as big as Tommy, would be disastrous.

One of the things that Noell does before he brings the gorilla out for a walk now is to move his car. He does not want any more $50 dents in it.

A Noell gorilla. Old pals together, Tommy and Robert Noell. (Photo by Norman Zeisloft.)

Tommy eats a number of things, including two bushels of fruit a day. Noell sometimes helps him with his food by peeling a banana and biting off some of it. As Tommy thrusts his black face into Noell's face, Noell will push the banana from his mouth into Tommy's mouth, which Noell has learned as being essential in establishing and maintaining a close gorilla-human relationship.

When Noell is tired of playing with Tommy, he usually chains him to a tree trunk for a time. This is often a signal for Tommy to rise up on his hind legs and slap his chest rapidly with his palms slightly cupped. He then grabs Spanish moss from the neighboring trees and throws it at the people watching him and finally bangs the ground with his fists. A number of dogs, puppies, and chickens are also in the compound with the gorilla. If a tiny young puppy comes near him, he will reach out and pick up the little creature. This causes terror in the onlookers, who feel that the pup is going to be squashed. Instead Tommy will get down on his forearms and put his nose down close to the puppy's and they will sniff each other. After a few moments the gorilla will begin a low grumble which seems to indicate that he is satisfied with the puppy, which hasn't suffered as much as a scratch, and returns it to Noell.

There are two more gorillas in the Noell's Ark ménage. Besides Tommy, there is Tarpie, a female gorilla who is eight years of age and weighs 300 pounds, and Otto, a young male of seven, who weighs 150 pounds.

The Noells are extremely fond of their three household gorillas and consider them their "children." It is possible that the Noells' gorillas are the only ones that are in a private home in the United States.

One thing about owning and raising a gorilla, the Noells point out, is that "when they want something they usually get it." On one occasion Tommy wanted Mrs. Noell to go to the other side of their backyard. When she didn't respond to his wishes, he got behind her, pointed her in the direction he wanted her to go, and then bit her on the bottom.

Gorillas, of course, inevitably become members of a circus menagerie, and the most famous of the circus gorillas was Gargantua, which belonged to the Ringling Brothers and Barnum & Bailey Circus. The unusual manner in which Gargantua was acquired by the circus has been described by Henry Ringling North.

In 1937 the future of the circus was in doubt and Henry

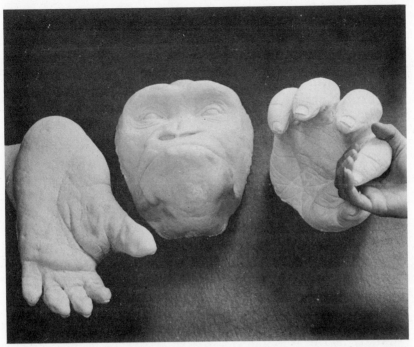

Death mask and hand and foot casts of Gargantua II. (Photo by Yerkes staff photographer Frank Kiernan.)

Ringling North, his brother, John, and his colleagues were staying at the Ritz Hotel in New York while they were engaged in very difficult decisions concerning the circus.

One evening the telephone rang, and the caller was a Mrs. Lintz, who claimed that she had in her possession a full-grown gorilla and would John Ringling North like to buy it for the circus? John replied, "I certainly would. When can I see it?" The lady seemed to become a little evasive, but then she invited them to go to her home in Brooklyn to have tea with her. John, who was hoping to get control of the circus, said that there was no doubt that they needed to have that gorilla. He went to Brooklyn in a taxi, which passed into a tenement district first and then into an elegant waterfront street. There at a brownstone mansion a small, middle-aged lady greeted them and sat them down on rosewood chairs to have afternoon tea with her. There was plenty of tea, and they drank gallons of it and chatted about everything under the sun until they began to feel that they had been the victims of one of the strange woman's fantasies. So finally they asked point-blank to see the gorilla. She said that it was certainly time for them to see it. She led them through her house down to the kitchen, into the basement, and then across a yard to a shed at the back where the horses had once been stabled. There was a naked electric bulb suspended from the rafters, and it seemed to illuminate mainly cobwebs and dust. Sitting on a wooden chair was the gorilla's keeper, Richard Kroner. At one end of the room was a box that looked for all the world like an oversized coffin. It was braced by heavy timbers, both from the ceiling and from the sides, and it also was lined with steel. Mrs. Lintz, before letting them see her gorilla, which she called Buddy, wanted to tell them his story.

Some years ago when this gorilla, at that time an infant, was traveling from Africa on a ship, a sadistic member of the crew, who had become upset about something the young gorilla did, threw acid in the poor baby's face, burning it horribly. Mrs. Lintz, a passenger, bought him from the captain, who thought the gorilla's appearance had been ruined and was glad to let go of him. Mrs. Lintz's husband, who was then living, had been a plastic surgeon. He had endeavored to carry out some repair work on Buddy's face, but without much success. Mrs. Lintz claimed that the young gorilla was very sweet, that he loved everybody, and that everybody loved him. He used to enjoy having tea with his mistress every afternoon. The front of the box in which he was then confined had a sliding door with slats in it. Behind the door were iron bars. When the door was raised for the circus people, a rather fearsome face leered out at them, twisted by the effects of the acid. The gorilla was eight years old, and very few gorillas at that time had survived that long in captivity. One of the Ringling family had actually imported two gorillas in the past, but they had survived for only six months. If Buddy had survived for eight years, it looked as though he had become acclimatized. Mrs. Lintz was determined that she would dispose of him, particularly since before they had strengthened the cage, the animal had escaped and found his way into Mrs. Lintz's bedroom. She woke up in the middle of the night to find him there. John Ringling North purchased the animal for $10,000. They decided that because he was a gorilla and also because his face was twisted by acid, he would be just about the most terrifying creature that the world had ever seen. They also decided to change his name and call him Gargantua. The next step was to get Gargantua down to Sarasota, where the circus was located, and to do this they hoped to get him onto a train called the Orange Blossom Special.

One of the family was familiar with the stationmaster at the

Pennsylvania Railroad, which ran the Orange Blossom Special. They asked him if he would take a gorilla down to Sarasota on that train when they traveled on it and got a negative reaction. Then, on the telephone, John Ringling North denied having asked the stationmaster to take a gorilla, saying that what he really wanted to take was a little delicate monkey that would have to go down with him because he would die of loneliness in a freight car. There was some resistance from the stationmaster since on one occasion a boa constrictor had got loose in a Pullman. After that the railroad had made a rule that dangerous animals were not permitted in baggage cars of passenger trains.

The party arrived early at the station with Gargantua's enormous box and Kroner, the keeper, with it. It was placed upright on an end in the baggage car. The stationmaster arrived to see how things were going and questioned the size of the box for such a small monkey. He was told that it was big because the monkey needed room to exercise. At that moment Gargantua did in fact exercise, and the box rocked on its foundations. The trainer said that the animal was probably feeling a chill and stuffed the end of the blanket in the small hole that had been cut in the front of the box to provide air. The blanket was suddenly whipped through the hole as if it had been a handkerchief. Then Gargantua straightened out the enormous nails that had been driven into the cage to hold the shutter in position and opened up the shutter. The end result was that Gargantua and the stationmaster were able to see each other, eye to eye. Fortunately at that moment the guard blew the whistle, and the train began to move. The stationmaster had to jump out on the platform. So Gargantua was on his way to Sarasota.

It was decided that Gargantua was very intelligent and that he was capable of thought. His mind was equated with that of "some maliciously capricious moron." Of course, he was quite playful. If anyone threw a softball into his cage, he would catch it and throw it back with an underhand throw. Then he would change his tactics suddenly and, instead of tossing the ball back, would throw it with great velocity at the head of the person playing with him. He also liked to play tug-of-war. If he was given one end of a rope, he would immediately pull it. He was known to pull it successfully with four or five men on the other end trying to pull against him. He would wind the rope under his arm to give him a little leverage, and then, with one pull of his arm, he would pull the four or five men off their feet. After he had won a tug-of-war, he would throw the rope out again for someone to pick up. Each time he threw it back the amount that he threw out was shorter, so that he would try to lure people playing with him closer and closer to his cage, where, presumably, if they came close enough, he would make a grab and bite them.

Before the circus went into action for the new season, the news of Gargantua was released. Immediately the press made him an international celebrity. One columnist even speculated as to whether he could beat heavyweight champion Gene Tunney in a fight. Gene Tunney, who was a friend of John Ringling North, played along and permitted the press to interview him, during which he said he would be happy to take Gargantua on, but he didn't go beyond just saying that he would do it.

Many people were interested in Gargantua. Dr. Robert Yerkes, founder of the Yerkes Primate Center, of course came to Sarasota to observe him. Also Dr. Bernard Baruch's brother, Dr. Sailong Baruch, arrived. He wore a beard and Gargantua was fascinated by the beard since he had never seen one before. He wandered around his cage, studying the strange creature with a beard from a number of angles. He finally decided that he didn't like men with beards and picked up whatever he could find lying loose in his cage and threw it at Dr. Baruch.

On one occasion he bit one of the Ringling brothers on the arm. He would not release the arm and had to be stopped from biting by Kroner, the keeper, who had to beat him about the head with a thick stick until he let his victim go. The bite was not too bad, but the headlines that went around the world were bloodcurdling, so much so that the German fiancée of John Ringling, Carlotta Gertz, rang him up one night from Berlin to inquire if he was dead. He was able to reassure her, however, that he was not.

The circus authorities were more and more concerned that Gargantua might catch from the circus crowds a germ which would cause his death, so they decided to build him a cage that would be air conditioned. Excited by this idea, John Ringling North, at 4:00 A.M. one morning, called up Lemuel Bulware of the Carrier Corporation, which was just getting started in the area of air conditioning. Mr. Bulware was angry at being called at that time of the morning, but John said to him, "You'll be delighted when you hear this. I want you to build an air-conditioned cage for Gargantua, great publicity for us both." Bulware was delighted with the possibilities and set to work straight away on the design of such a cage. In actual fact, what was built eventually was a cage wagon on wheels. The Carrier Corporation air-conditioned it so that it reproduced exactly the climatic conditions in the Congo. It had not only thermostatic control but also humidifiers. The term "jungle-conditioned cage" made quite a publicity phrase for the circus. Despite the air-conditioned cage, however, Gargantua got pneumonia once when he was appearing at Madison Square Garden in New York. He had the best doctors in the United States to treat him, and they managed to pull him through.

At his prime Gargantua weighed 550 pounds and was 5 feet 7½ inches tall. His arms had a total span of more than nine feet. He was not fed meat, but in addition to fruit and vegetables, he got boiled liver and also cod-liver oil to supply him with vitamins. In 1941 John Ringling North heard that Mrs. Stephen Hoyt of Havana had a female gorilla for sale whose name was Toto. North paid a visit to Havana and found that Toto was living in her own handsome small house in the grounds around Mrs. Hoyt's mansion. The final straw that had broken the camel's back was that Toto, while having tea in the garden, had accidentally broken both of Mrs. Hoyt's wrists, and that was the end as far as Mrs. Hoyt was concerned. Toto was purchased by North, and everybody had a wonderful idea of a marriage between Gargantua and Toto. The wedding was planned for Washington's Birthday in 1941, and a cake was flown down from Schrafft's in New York. Mrs. Hoyt was matron of honor and was clad in flowing chiffon, complete with a picture hat. Many reporters were asked along for the ceremony. They played a wedding processional as Toto's white cage was brought in, and on it was a placard that said MRS. GARGANTUA, THE GREAT. Jose Tomas, Toto's keeper, who had come with her, was actually riding in the cage with her so that she would not be nervous. It was backed onto the end of Gargantua's cage, and that end was taken away. Gargantua was absolutely amazed by what he then saw, since this was obviously the first gorilla he had seen since he was a tiny baby, and came forward, grasping the bars. Toto at first had her back toward him, but when she looked around and saw him, she panicked and rushed to her trainer and flung her arms around his neck. Then she turned and vocalized in no uncertain fashion at her would-be husband. That sent them both into a rage. They both rushed around their respective cages, roaring at the top of their voices, and Gargantua picked up all the half-eaten fruits and vegetables in his cage and pelted his prospective wife with them. While the congregation was not quite sure whether it should scream with laughter or run away in terror, the marriage was never consummated. Later they hoped that putting the cages close to each other and having the animals separated only by the bars of their cages would help them to get used to each other. Toto did not seem too averse to this possibility. As North said in his book, "She made coy advances, like throwing an overripe melon at her husband. But Gargantua spurned her. George Jean Nathan always claimed he was a 'fairy.'"

Another Gargantua anecdote concerned Mrs. Hoyt. She used to come from time to time to see Toto. On the way she would pass by Gargantua's cage. One day, without thinking, she went too close to Gargantua's cage, and as she passed, he grabbed at her but managed to get only her dress. There was just one rip, and Mrs. Hoyt stood there clad only in her bra and panties and, according to North, "screaming bloody murder." It was not the first time she had been treated this way by a gorilla, since this was a relatively frequent experience with Toto when she lived in Havana.

Gargantua's coat turned silver when he was twenty-one, and he got progressively more feeble. In 1949 the circus was in Miami, and that was where Gargantua died. North had promised to send Gargantua's body to Dr. Yerkes at the center for an autopsy, and he did just that. After he had been autopsied, the skeleton was prepared and sent to New Haven. As North said, "Harvard wanted him, but Gargantua was a Yale man."

Later on the circus obtained another gorilla, Gargantua II, who died recently and was autopsied at the Yerkes Primate Center.

4

They Make Lovable Pets—Almost!

PEOPLE get medals for performing brave and sometimes foolish deeds on the field of battle. There ought to be a special award of valor (and foolhardiness) for those who brave the trials and frustrations of raising a gorilla as a household pet.

Keeping a pet gorilla in your living room would be just as destructive as keeping a dinosaur (if you could get one). Weighing the advantages, a gorilla is lovable and will love you, in his or her starved need for affection, almost or perhaps actually to death. They make incredibly good watchdogs. There has never been a record of a burglar breaking into or entering a home with a gorilla occupant. They are extremely strong and can be trained to move or lift heavy furniture or automobiles that would require several men. On the other hand, the disadvantages are numerous. First of all, they are expensive. When last available on the open market, they sold from $5,000 to $15,000 apiece. They are extremely susceptible to colds and respiratory infections. They eat, depending on their age and size, from twenty to forty pounds of fruits and vegetables a day. Some develop a taste for exotic and costly food and become quite irritable when they don't get it. We have spoken of their destructiveness, but it bears mentioning again. A spoiled or frustrated gorilla is like a child. In a temper tantrum the upset ape can stove in the refrigerator, washing machine, and half the furniture with a few brawny swipes of his massive paw. With few exceptions, the gorilla pet and its owner eventually can't cope with each other and the big ape is shipped off to a zoo.

One of the bittersweet experiences in raising a gorilla was told by Mrs. Maria Hoyt, whose life was enriched by beautiful, as well as tragic, moments in raising her gorilla, Toto.

In the earlier part of the century Mrs. Hoyt's husband was hunting and collecting animals in Africa. Mrs. Hoyt had accompanied him and on one occasion was summoned to a hunt where a group of gorillas had been disturbed. Her husband had planned to kill only one gorilla, but the Africans, anxious for the meat, ambushed all of them and killed most of the group. Mrs. Hoyt found a tiny baby weighing only a few pounds and still clinging to its dead mother. She picked it up and took it to her camp. Then she made a little crib in her tent in which the animal slept. The baby was christened Toto, and for a time Mrs. Hoyt was worried about its survival because she could not get it to eat anything. Eventually she found an African woman who had more milk than her own child could take, and she suckled Toto. There are quite a number of observations of this type of human-ape relationship recorded in the literature on apes. Toto learned all kinds of things very quickly, including how to wipe her face with a handkerchief and after a while how to blow her nose in it. Eventually the Hoyts had to leave the French Congo, and they decided to take the little gorilla with them in a deluxe suite on the steamer which took them from Africa to Europe; Paris was their destination. In the suite on the way to Paris, Toto quickly learned that she could come in each morning from the room where she slept and get into bed with Mrs. Hoyt, and together they would have breakfast. She soon turned this into a habit, and each day of the voyage she would turn up and go through the routine. When they arrived in Paris, they had accommodations in a hotel in the Rue de Rivoli. Presumably Toto felt very insecure under these circumstances and would yell at the top of her lungs if left alone; she always slept in the same bed with either Mrs. Hoyt or her husband. It was not long before the chill air of Paris gave her a bad cold. The best Paris pediatrician was brought in to attend her, but in spite of the care and attention, she developed pneumonia, and Mrs. Hoyt went through a traumatic vigil of eight days and nights trying to pull Toto through a crisis. During the course of this illness they even went to the expense of bringing a portable X-ray machine into the hotel room to take pictures of Toto's lungs. Toto eventually recovered from the disease but was very weak for a long time, so the doctor recommended that the Hoyts take her out of Paris to the sea. They decided therefore to go to a small town near Bordeaux. The retinue which they formed on their departure must have caused great amusement and interest to Parisians, who crowded around to see them leave their hotel. The retinue consisted of Mrs. Hoyt and her mother; Abdulla, a tall, handsome Swahili; their chauffeur; two pomeranian dogs, and Toto.

When they got to their destination, they received a great welcome. Toto became the pet of everybody, the honored guest, and became very spoiled. In fact, she would not go to sleep at night unless Abdulla sat on her bed; often she would pretend to sleep, and as soon as Abdulla got up and began to tiptoe out of the room, she would sit up and scream. Mrs. Hoyt noticed at this time that Toto had a very keen sense of hearing. Toto could hear Abdulla walking on the tile floor of a neighboring bathroom in his bare feet.

During Toto's convalescence Mrs. Hoyt found that one of the problems of owning a pet gorilla was that whenever she wore bracelets, necklaces, or even earrings, Toto tended to tear them to pieces. Sometimes she would run up to Mrs. Hoyt when she was not looking and grab the necklace from around her neck, break it, and run off with it. On many occasions the Hoyts spent long periods on their hands and knees trying to retrieve pearls that had been dropped all over the room in this way. Toto was cured of this behavior by being given a bracelet and necklace of her own.

Eventually Toto recovered her health and the Hoyts decided to leave France and seek a climate more favorable to their gorilla pet, and it was decided that Havana would be the ideal place in which to settle. Mrs. Hoyt and her entourage boarded a little launch at Cherbourg headed for the transatlantic liner which was to take them to Cuba. Toto had been given a sleeping pill before they carried her aboard, wrapped up like a human baby, and taken to the Hoyts' suite of rooms. Mrs. Hoyt's pomeranian dogs were also carried on in a disguised way, so that no one on the ship knew that these animals were in the suite. Throughout the journey the Hoyts had everything organized so that whenever the stewards were in any of the rooms, the animals were in another one, so the stewards themselves never even knew the animals were there.

One of the problems with Toto during this period on the ship was that she would try to climb out of the porthole whenever it was open. She also made strong efforts to catch the blades of the electric fan. In spite of great care, Toto managed to break up several chairs. Then she tore to pieces most of the curtains in the suite. These they managed to conceal until the last day of the voyage, and Mrs. Hoyt went to the captain of the ship and told him about their wrongdoings and paid him for the destruction.

Mrs. Hoyt's husband, who had reached Havana before they arrived, met the ship at the dock, accompanied by members of the press, who were all very pleased when they saw Toto dressed up in baby clothes. She was the object of many pictures and attempted interviews on that day. Toto was excited and happy when Mr. Hoyt called out to her, and she threw her arms around his neck. It was a long time before she would let him go. In the hotel where they stayed there was a double bed in one room, and in the adjacent room was the bed in which Toto slept soundly and relaxed as long as the communicating door between the two rooms was left open. Eventually the Hoyts found a house to rent in the country-club park in Havana. It had extensive grounds and gardens, which Toto took to instantly. Toto was nine months old. This was the time when she stood upright for the first time, and Mrs. Hoyt was called excitedly by Abdulla to come and see Toto walking while holding his hand. Toto was delighted with herself, murmuring *whu whu* all the time, and kissed Mrs. Hoyt repeatedly. Once she had made a start at walking, she soon was seen walking with Abdulla all over the grounds. As Toto began cutting her teeth, her mistress made a big fuss of each tooth that appeared. Toto soon got the idea that they were important and used to stand in front of a long looking glass opening her mouth and gazing at the little teeth sticking out from it. Presumably, when she did this, she knew it was her image, yet there were occasions when she would show anger at the figure in the mirror, attack it, and attempt to destroy it.

Toto took rapidly to picture magazines and spent many hours looking at them, turning the pages and not tearing them as many of the Yerkes apes will do. Bobby, our chimpanzee, and some of our other animals have been supplied with picture books from time to time, but they look at the pictures for only a little while and then rapidly destroy them and the book. When Toto was supplied with newspapers, she would dance on them and would love to crumple them up; she was so pleased with the noise they made when they were stamped on, and she also liked the noise of tearing them. A tree ocelot, Cleo, which used to live in the Bournes' house, also had a weakness for tearing papers. She did this partly to attract our attention, partly because she apparently liked the noise.

There was a swimming pool on the grounds of the house where she lived, and when Mrs. Hoyt went swimming, Toto would panic, scream with fear, and run around the edge of the pool in excitement, waving and leaning over. When Mrs. Hoyt came near her, she would grab her bathing cap and try to drag her out of the pool. She behaved as if she were frightened that something might happen to her mistress in the water. This is not very surprising because the great apes all have a fear of deep water and do not try to swim if they fall in water that is over their heads. During the summer that the Yerkes Center had the chimpanzee Bobby and a little female chimpanzee, Jenera, on an island in a nearby lake, they would go into the water on many occasions up about waist high, but they never made any attempt to go any deeper or to swim away from their island. There are some records in some of the continental zoos of gorillas and orangutans drowning by falling into the moat surrounding their dens. We also lost a gorilla in this way from a group which we had placed temporarily on an island in a lake in a neighboring African safari park. The animal ran into the water and found itself out of its depth and drowned before it was possible to rescue it.

Among the various presents given Toto, some had special interest for her and others she ignored or destroyed. For example, she was given a toy elephant, a toy teddy bear, and a squeaky ball. These she either threw away or tore to pieces. Then they gave her a spinning top, hard rubber balls, and white chalk. She very quickly learned to spin the top with her fingers and needed only one lesson before she became an expert at spinning it. She bounced the hard rubber balls on the floor and caught them, and she would sit for hours on the stone path, scribbling with the chalk on the stones.

At about this time, 1930, Madame Rosalia Abreu, who had a private colony of primates in Havana, died, and her colony was redistributed. Her animal handler became available to the Hoyts, and they employed him to look after Toto. This Cuban animal keeper, Tomas, used to watch Toto drawing, and whenever she would do anything that appeared to be a number, he would play it in the lottery; however, he never won. Sometimes Tomas, the gardeners, and the butler would all be seen gathered around Toto deciding what the number was that Toto had just written so they would bet it in the lottery.

With the arrival of Tomas, Abdulla went back to Africa. He had come to Havana in a pair of khaki shorts and a shirt, and he returned to Africa with several trunkloads of ties, shirts, underwear, shoes and hats, etc. Mrs. Hoyt said that in his smart blue business suit he looked more like a prosperous Harlem bandleader than a native African. When he got back to Tanganyika, he had enough money to start the first of a chain of mercantile stores.

Madame Abreu, the previous employer of Tomas, had a remarkable collection of primates. She started with a macaque monkey and eventually became interested in anthropoid apes. Later she accumulated a great collection of primates, which were described in *The Ape People*. Madame Abreu was a very religious woman and, according to Mrs. Hoyt, was convinced that apes had immortal souls; she even had her apes attend mass with her in a private chapel erected on her own estate. One of Madame Abreu's guests described how she went into an enclosure with two chimpanzees carrying a rosary. One of the chimpanzees cowered when it saw the cross and retired to the rear of the enclosure, whereas the other came forward and appeared to kiss the cross. Madame Abreu said of the chimp which cowered, "He has been evil and knows it, but dear little Clochette is good."

Tomas took complete charge of Toto and was with her day and night for more than six years, even sleeping in the same bed in a room assigned to the two of them. The Hoyts later made a special bed for Toto. In this bed a hole was cut through the mattress and springs and under it a vessel was placed; Toto used this for her toilet purposes at night, but in the daytime she used a conventional toilet. She also learned to turn on all the faucets in the new house and the garden and, of course, produced floods every now and again. Tomas trained Toto to sit at the table and eat with him.

Problems of keeping an anthropoid ape in the home now began to bear down on Mrs. Hoyt because by the time Toto was three years old she was as strong as two men and had the ingenuity of a dozen growing boys. She could open all the doors whether they were locked or not. She knew how to find keys, and if she could not open the doors with the keys, she would either break the handle of the door or knock the door down. She learned that she could crash through a door by simply putting her shoulder against it and pushing, so it became impossible to keep her in one room simply by locking the doors. Another of her tricks was to slide down the banisters, and the

Hoyts tried very hard to break her of this habit. Her other activities made the Hoyts realize that it would be extremely difficult to keep her much longer in the house. One of her activities was moving her heavy iron bed across the room and getting close to the wall. From there she could reach the bell to the servants' quarters. She would then set it clanging at any hour of the day or night and produce chaos and panic in everyone from the Hoyts, to the butler, to the maid.

Finally the Hoyts decided to build a house for their mischievous animal. The house had a living room fifteen feet by twenty-five feet, a bedroom ten by fifteen, and an outside enclosure forty by eighty, which was surrounded and covered by heavy iron bars. In it they put benches and swings and various play equipment which Toto could use, and usually the door to the outside enclosure was left open so that she could go in and out of the gardens at will. Generally they locked her in the area only when they had visitors. Even in this sophisticated retreat and with her sophisticated upbringing, Toto would build herself a nest of palm leaves on which she would lie on hot afternoons. By this time Tomas had finally managed to extricate himself from having to sleep with her at night and she became accustomed gradually to sleeping by herself. The Hoyts were greatly impressed by the extreme intelligence of Toto; but they pointed out that although she was highly intelligent, she was not intelligent enough to learn human self-discipline, and that was really the factor that made it impossible to keep her in the house. Toto's life, unfortunately, fell between the two planes of being neither a human nor a dumb brute, and she lived in what the Hoyts described as a half-world to which she had been conditioned by their companionship. This is the problem that exists with animals kept as pets that become too big and too strong for their owners and eventually have to be sent into what for them must seem like a penitentiary for the rest of their lives. That is to say, they have to be sent to a zoo. As a result of this half-world that Toto lived in, she wanted to share human life to the fullest, but she did not understand why limitations should be imposed on her.

This was a period when Toto adopted a kitten called Blanquita. She would pick it up very gently, cuddling it in her arms. The kitten was apparently as pleased as Toto with this treatment, and it snuggled against Toto and purred very happily. After that the cat was constantly with the gorilla, and wherever Toto went she carried the cat; sometimes she carried it under her arms, sometimes just clinging to her back as a baby gorilla does when its mother is walking or climbing, and sometimes she wore it wrapped around her neck like a fur stole. The cat became similarly dedicated to Toto. Eventually Blanquita grew up, and one morning she produced a litter of six kittens in Toto's bedroom. This seemed to upset Toto, and she looked and behaved as if she were angry. She walked over to the new family of kittens, looked them over, and finally picked one out. This was a black and white kitten rather like its mother in appearance. From the moment she adopted the kitten she behaved as if the kitten's mother did not exist, and after having been Toto's loving companion for so long, Blanquita found herself on the outside. She would come and rub against Toto and purr, but Toto would not pay attention to her at all and would even brush her off, demonstrating quite plainly that she did not want anything to do with her. She bestowed all her attention on the baby, which was named Principe by the Hoyts.

Toto was very fond of Mrs. Hoyt's mother and would sit for long periods stroking her white hair. Mrs. Hoyt's mother taught Toto how to make pictures in sand with her fingers. She would draw a circle to outline the face and put three dots inside, two for the eyes and one for the nose, and she would put a straight line for the mouth. Often Toto was seen to greet the old lady by making this drawing in front of her in the air as she approached. Later Toto actually drew this representation of the face on the flagstones, but whether she recognized it as a face or whether it was simply a series of movements which she had learned, it is difficult to say. Toto was also a great eater of mosquitoes and ants. If she saw a group of ants, she would brush her arm against them so that a number of them would stick to her fur, and then she would lick them off just as if she had been in the wild. It is of interest that most of the observers of gorillas in the wild have said that they saw no sign of gorillas eating any kind of animal material, either insects or flesh or birds or eggs. Toto's reaction, however, does suggest that they may in fact eat insects and that the observers have just not seen them doing this. The diet which the gorilla is said to eat in the wild would be grossly deficient in vitamin B_{12}. This vitamin would be essential for the animal to remain as well and healthy as, in most cases, it seems to be. So it must have some source of vitamin B_{12}, and it is very likely that insects could provide this.

Toto was very fond of Mr. Hoyt, but she had no respect for

his straw hats or neckties. She was especially fond of his ties, and if he came near her wearing a brightly colored one, she would suddenly grab at it and with a quick jerk rip it from his neck. Sometimes when the tie did not break, the force of the pull would throw Mr. Hoyt on his face. On another.occasion a friend of the family who was a very strong man and a boxer came to visit. He was warned about Toto's strength before he shook hands with her, but scorned any danger. The moment Toto got her hands on his, however, she gave him one pull and threw him against the bars of the enclosure with such force that it seemed that he might have broken a shoulder; fortunately, however, only a muscle was bruised.

Toto sometimes played a mischievous trick which caused some concern to the gardeners. To trim the tops of the tall trees, the gardeners had to climb up a ladder leaning against the trees. If Toto was loose and saw a gardener climbing up the ladder, she would immediately chase after him and run up the ladder behind him. The gardener would be so terrified that he would go up to the top of the ladder, over the tree, scramble or fall down the other side, and rush off at high speed. Toto would then jump off the ladder and chase him madly for a little while. On these occasions she would then pick up the pruning shears the gardener had dropped and throw them after him. It all sounds very much like a comedy sequence from an early movie.

Toto's meals kept a cook busy the entire day. At 7:30 A.M. she had yogurt mixed with sugar water. At 8:15 A.M. she was given a quart and a half of milk and tapioca and at 10:00 a large mug of orange juice. At 11:30 she was given eight bananas and some cream cheese. At 1:00 P.M. she had lunch on a tray. At first she was given no meat with this, but since she ate with Tomas, she soon began to steal the steak and chicken from his tray and he insisted that she be given the same type of food. At 4:00 P.M. she had a quart and a half of milk with oatmeal or pablum or coca malt. At 6:00 P.M. she was given two baked apples and drank a jug of milk. Between meals, of course, she had water ad lib. She had no problem indicating to people that she wanted a drink of water. She would look at the water faucet, indicate it with a finger or a hand, and smack her lips rapidly. The articles of food included from day to day are of interest: vegetable soup, fresh carrots, potatoes, sweet potatoes, rice, barley, tapioca, eggs (scrambled or boiled or served as an omelet), beefsteak, lemon meringue pie, apple pie, and fruit compote. Her favorite fruit was peaches, but she also liked grapes, oranges, and bananas. She did not like melons at all. Although she loved baked apples, she would only nibble at a raw apple.

Whenever Toto had reached the violent stage, Tomas had to use an electric cattle prodder to control her. He would also show her a snake which he kept imprisoned in a bag attached to his waist. These two were valuable forms of control during a difficult period.

The remarkable strength of this animal was shown by the fact that on one occasion when she went into the garage and found the station wagon in her way, she took it by the rear axle with one hand, and even though the emergency brake had been put on, she pushed the car against the wall of the garage with such force as to smash the headlights. Even when she was walking with Mrs. Hoyt, there were times when she would unpredictably grab her dress, give it a jerk, and either tear off a large piece of it or else pull her onto the ground. Sometimes she would stand erect and then charge, and Mrs. Hoyt said the only way not to be hurt when this happened was to fall to the ground and roll out of her path. Sometimes, however, Toto would seize her by the arm or even by a portion of her dress and pull her along the lawn or walk, often releasing her only after cries had brought her assistance. Even when Mrs. Hoyt was playing with her in quite a gentle way, kissing her and putting her arms around her, she would every now and again on sudden impulse rip off Mrs. Hoyt's dress and tear it to pieces. Because of this, her mistress always had a special locker in Toto's playroom, and in it she kept a variety of clothes so that she would have something to put on when she left the animal playroom. At about this time Mrs. Hoyt found that the delivery boys who were bringing her groceries refused to come to the house and simply threw their deliveries over the wall onto the grounds, which caused considerable inconvenience to the household. Eventually it was decided that the complications of keeping such a strong animal in a home were too much. All those associated with Toto were in danger not because the animal was vicious, but because she did not know her own strength. She was potentially dangerous to all the neighbors. Mrs. Hoyt found herself forced to make a decision to give Toto to Ringling Brothers Circus, which already had a large male gorilla, Gargantua. Principe, the cat, had accompanied Toto on the expedition to the circus for she would have been very upset if the cat had been left behind. Unfortunately in Boston when the circus was traveling, Principe was lost, and the circus was forced to move on before she was found. In Chicago Mrs. Hoyt discovered a cat who resembled Principe and she hoped that Toto would accept her in lieu of her pet. But Toto would have nothing to do with her and pushed her away. Toto remained with the circus until 1956, at which time Mrs. Hoyt persuaded the circus to sell Toto back to her and the animal died late in 1968.

We have dealt with Mrs. Hoyt's experiences with Toto at some length because they demonstrate two points. One is that however intimately you are associated with a young ape, when it grows up, it is difficult to keep in the home. The animal just does not understand the restrictions that such an unnatural existence places on it, and it will use its great strength to get rid of any that are unpleasant. The second is that Mrs. Hoyt's detailed description of her ape brings out many remarkable aspects of the animal's intelligence, especially its ability to draw a representation of the human face. These descriptions help to underline the fact that the great apes have an intelligence which we have scarcely penetrated. This is probably owing to a failure to communicate adequately with them. We will talk more about communication in a later chapter of this book.

Another newsworthy gorilla which was kept in a private home was John Daniel. He lived with Alyse Cunningham in London between 1918 and 1921. Just as Toto had thrived on affection when she was with Mrs. Hoyt, so John Daniel thrived on the affection of Miss Cunningham. John Daniel was described by his mistress as "very much afraid of full-grown sheep, cows and horses, but he loved colts, calves and lambs, proving to us that he recognized youth." It is interesting that the more intelligent large mammals seem to have an intuitive gentleness toward the young, even of other species. Colin Groves, an anthropologist and an expert on gorillas, points out that humans, for instance, have a kind of intuitive desire to mother young animals and that some breeds of dogs, for example, Pekingese and the Chihuahua, "have been deliberately bred to look infantile and so retain the motherable quality throughout life." Gorillas are emotional and react to fear much as humans do, except perhaps a little more violently. They can, however, be very gentle with people whom they have a strong affection for and also seem to have some degree of ability in abstract thought and reasoning.

Alyse Cunningham wrote about her life with her pet gorilla, John Daniel, in the *Zoological Society Bulletin* (Vol. 24, 1921).

John Daniel was one of the most famous pet gorillas and held court in London in the early 1920's. John had been brought in with a group of monkeys which had been required for some experimental studies by the British government. He had been captured at the age of three years in the French Gabon area of West Africa and was acquired by the captain of a French ship, who had brought him to Le Havre. There an animal dealer bought him and sent him to England with the monkeys. After arrival in England, he was sold to a London department store, in which he was used as a Christmas attraction. Alyse Cunningham was not oriented toward monkeys at that time and did not take particularly to the gorilla, but she thought he was very entertaining in his cage, where he would beat his chest and clap his hands or do little tricks which he seemed to understand would amuse the crowds. Her nephew, Major Rupert Penny, had always had an interest in primates, and he decided to buy the gorilla because he was interested in seeing the level of mentality of an animal which was so close to humans. John Daniel amused people in the zoo section of the department store until very close to Christmas. Then he acquired a severe cold or possibly even influenza, which he almost certainly caught from the public to which he was exposed. He had been in a very stuffy atmosphere with Christmas crowds during the whole day and kept in close proximity to people, and the temperature night and day was around 80 to 85 degrees. On being purchased by Major Penny, John came to live with the Cunningham family at 15 Sloane Street, at that time a very smart section of London. When he came there at the end of December, he was very rickety and weighed only thirty-two pounds and had difficulty in standing erect.

The family converted a small room in the house into a cage for him. It was separated from another room by bars so that he could see and be near people all day. The cage was heated by an electric radiator, but the windows had been taken out of the cage in one part and the space covered with a number of thicknesses of muslin so that fresh air could get into the room. The high temperature at which he had been living was dropped to 65 degrees, and his cold got better immediately.

It was impossible to leave him alone at night because he became frightened and lonely and would shriek continuously until someone returned to the house. The new owners found on inquiry that he had reacted that way in the store also before he came to live with them. He would begin to cry as soon as he saw the assistants putting things away prior to closing the store for the night. This loneliness was very trying on his nerves and on his appetite. Major Penny very soon had his bed made up in the room next to the cage, and John responded immediately by being happier and putting on weight.

By this time Miss Cunningham began to grow fond of the gorilla and became interested in him. She fed him and washed his hands and face and his feet twice a day and even brushed and combed his hair. If he got hold of a brush or comb, he would try to do it himself. He soon found this routine very enjoyable. The next bit of training was to try to get him to be clean in his bowel habits. They had great trouble in getting John toilet-trained even though he was petted and offered grapes, which he liked very much. In the beginning when they tried to get him to go to the toilet, he would roll on the floor and shriek, but eventually he understood what he was expected to do and he behaved very well. It took them about six weeks to train him in this way. About February, 1919, he was allowed out of his cage and was given the freedom of the house. Under those circumstances when he had the urge, he would go upstairs to the bathroom, turn the doorknob, and go in and use the bathroom as a human does.

John had a nongorilla appetite. He would try all kinds of foods, but he tired of them very quickly. Milk, however, was one of his favorites, and he always liked to drink it warm. He started off with a quart a day and later on consumed three and a half quarts. He liked to choose his own foods, and they would give him a selection of several kinds; he was especially fond of bananas, oranges, apples, grapes, raisins, currants, dates, raspberries, and strawberries when they were in season. He liked all of them to be warmed. He would never eat anything that was stale. He was very fond of jelly, especially fresh lemon jelly, but he would eat it only on the day that it was first made. He enjoyed eating roses more than anything else, and the more beautiful the rose, the better he enjoyed it. Faded roses did not tempt him at all. As a result of this, Miss Cunningham was never able to have roses in vases inside the house.

John's fame soon brought numerous curiosity seekers to the house on Sloane Street to see the animal described by Miss Cunningham as an "ape child." In the beginning they used to let in everyone who came to the door to see him, but after a while people became so numerous it was impossible to let them all in, and they had to reduce the visiting hours and visitors to those introduced by their personal friends or people who had a scientific reason for seeing the animal. When he was in the presence of people who came to see him, John would entertain them just like a self-conscious child. He would take the visitor by the hand and lead him around the room. He was very quick to note that the person was nervous, and John would then make a run past him and smack him on the leg as he went past. As Alyse Cunningham said, "You could practically see him grin as he did this." Another game he was very keen on was to shut his eyes and then run about the room, running up against tables and chairs. He also liked to empty the contents of a wastepaper basket all around the room. Then he seemed to enjoy picking everything up again and putting it back in the basket, but only when he was told to. While he was doing it, he would have a very bored expression on his face. When the basket got very full, he would push the contents down so that he could add more to it. If he removed an object from its proper place in the room, he would always put it back in the right place when he was told to. This applied even to books on the bookshelf. He had very good table manners and always ate at the table with the rest of the family. As soon as he realized a meal was ready, he would get his own chair and pull it into place at the table. He did not eat a great quantity of food, but he was especially partial to drinking water out of a tumbler. He rarely ate bread, but would often eat butter. He enjoyed very much afternoon tea, which consisted of thin pieces of bread with plenty of jam on it. After dinner he liked a cup of coffee. He showed no signs of greediness, which many other animals demonstrate, and he was never seen to snatch at things.

Gorillas in the wild have not been seen to drink, but John drank a lot of water with his meals and would sometimes even get water for himself by turning on the tap. He always turned the tap off when he had drunk as much water as he wished to.

The residence in Sloane Street was a town house and John lived on the upper floor; if he found a window unfastened, he would open it and lean out. When a big crowd had collected outside, he would then beat his chest with his hands. If he saw children in the crowd, he would put his hand out toward them and make grunts which seemed to be of pleasure. If he was told to stop what he was doing and come inside and shut the window, he would do so, but in addition to shutting the window, he would always lock it. Miss Cunningham had a little niece, three years old, and sometimes she would come to stay in the house with her mother. John Daniel and she seemed to understand each other very well and would play together by the hour. If she got upset about anything and cried, her mother would go to her and pick her up. On such occasions the gorilla would make an effort to give the mother a nip or even smack her hand with his hands. In some way he seemed to blame the mother for the child's crying.

One day as Miss Cunningham and her sister were planning to go out, the gorilla decided he wanted to sit on Miss Cunningham's lap. Her sister said, "Don't let him, he will spoil your dress." She did, in fact, have a light dress on. There is no doubt that the gorilla would have spoiled it, so she said, "No." The gorilla lay on his back on the floor and cried like a child. Then he stopped, got up, explored the room. Eventually he discovered a newspaper, which he picked up, spread on Miss Cunningham's lap, and climbed up onto her lap, sitting on the newspaper.

Bedtime was eight o'clock at night. He had his own room, which had a bed with a spring mattress and blankets. If he got up in the night, when he got back to bed, he would pull the blankets over himself, just like a human. He liked to stand on the top rail of the bed and leap onto the bed, which would tumble him over because of its springiness. There was never any attempt to teach him tricks. He was taken by train to the cottage which they had in the country and would travel just like an ordinary passenger; he didn't even have a chain around his neck for these journeys.

Out in the country he showed no liking for open fields or any kind of open country, but in a garden where there were plenty of hedges and large plants he was much happier, and he was very much at home in the woods.

The family found that as John grew older, he was becoming more and more of a challenge to them. They tried hard to get someone that they could employ to look after him, but everyone they tried was under the impression that if you were bringing up an animal, you had to do it with a stick, but a gorilla does not put up with these methods from anyone. Eventually the only way they were able to punish him was to tell him he was naughty and push him away. He would be very upset by this and would roll around on the floor and cry and give every indication of being repentant. He would, for instance, take the ankles of the person who had pushed him away and lift them up until the feet were on his head.

Finally it was decided that life was too complicated with him,

John Aspinall plays with one of his pet gorillas. (Photo by Harry Zeyn.)

and John's owners decided to sell him on the understanding that he would go to a private park in Florida. They thought that life in Florida would provide an ideal situation for him, and they signed the sales contract on the understanding that the man who was to take him away would stay six weeks so that the animal would get to know him. When he actually turned up, he was prepared to stay only a few hours, and so the poor gorilla was forced to leave with a man who was a total stranger to him. He soon became homesick and eventually very ill. Miss Cunningham was cabled to come to New York as quickly as she could, but he was dead when she arrived. His place of death was the Madison Square Garden Tower in New York. He died at the end of April, 1921.

Two of John Daniel's activities were hilarious. On one occasion he got out of the flat and flagged down a taxi on Sloane Street. This was done in such a human fashion that the taxi driver actually pulled up and opened the door for him to get in. Another time, when the guards were marching to go on duty at Buckingham Palace, John Daniel appeared and carried out such peculiar antics that he caused the guards to break step and lose the typical composure for which they are famous.

John Aspinall and his gorilla family enjoy a picnic on the grounds of his estate near Canterbury in Kent, England. (Photo by Harry Zeyn.)

Intimate conversation between John Aspinall and one of his gorillas. (Photo by Harry Zeyn.)

Near the ancient cathedral city of Canterbury, situated in the county of Kent, southeast of London, there is a grand old estate which is owned by a businessman who has a great love for animals. Among the animals that he likes are gorillas, and he owns eight of them. The gorillas have their own house, especially designed by an English architect who spent some time doing research with Dr. Ernst Lang in Basel to learn gorilla living habits. The building is a roomy one with heating to take care of the animals in the cold weather. Adjoining it is a large cage where the animals can sit in the sun and enjoy fresh air, if it is not too cold, during the daylight hours.

The man who owns these gorillas, John Aspinall, is reputed to have made huge gambling coups which have provided him with what has been described as a princely standard of living. Aspinall said that he had once been asked why he bothered to get gorillas for his collection of wild animals. For the same amount of money he could get a breeding pair of tigers or even a whole herd of Asiatic deer or perhaps even a white rhinoceros. His reply was that the gorilla is the star of any collection of animals. He said, "The secret of their fascination is their physical stature combined with the deep impression made by their personalities. They dominate any ape house. Where the orang is weird, the chimp farcical, the gorilla is arresting. The immediate response to most humans on confronting a full-grown gorilla is one of awe."

From time to time the animals are allowed out to stroll in the park with Aspinall and his family. They have a gay time galloping into the paddocks and are not at all fazed by the nilgai antelopes and the deer which Aspinall also keeps there. There were some fine big beech trees in the park which the gorillas loved to climb. They also like to eat some of the fresh foliage. When Dr. Lang, the director of the Basel Zoo, visited the park and was out with the gorillas and their owner, he was standing, watching them, and suddenly received a sharp blow. He managed to stagger a few yards away and recognized the donor of the blow as Gugis, a gorilla who at that time was six years old and had considered himself to be the chief of the apes. He had run up behind Dr. Lang and whacked him. The gorillas do not do this to their owner, whom they seem to regard as the top boss, but test out any other male with him by attempting some kind of semiviolent contact of this sort. Passersby can sometimes observe Aspinall's gorillas through a wire fence, and when they gather to watch, the gorillas often wander over to take a look at the people staring at them. Although they would have no difficulty in climbing over and startling the visitors, they make no attempt to do so apparently being content to be restrained by the fence and just stare back at the people outside. The young male gorilla, during Dr. Lang's visit, continued to try to pick a quarrel, and Dr. Lang had to keep a close eye on him. The gorillas had a sand pit where they enjoyed making sand pies. Their method is to fill a little bucket with sand and tip it upside down until they have two or three of them around. Then with one sweep they demolish them. Even when their owner or others made sand pies, they seemed to resent the sight of them standing there in a neat orderly shape, and they would rapidly destroy them.

One of the things noted with Aspinall's gorillas is that when they were out walking in the paddocks and it began to rain, the gorillas didn't mind at all; they seemed to enjoy it and would roll around in the wet grass. This was also noticed in the San Diego Zoo gorillas, particularly the big ones, M'bongo and N'gagi. When it was raining they would romp and tumble around in it and slide about—even when it was pelting hard. Sometimes they would even refuse to go to their sleeping quarters for the night when it was raining, for, like children, they would much rather stay out and play in the rain all night.

One of John Aspinall's gorillas is a little mountain gorilla called Kivu which came to his house some years ago. He was three years old when he came, and he weighed forty-two pounds. He was looked after by Dorothy Hastings, Aspinall's mother-in-law. He was very sick when he arrived, coughing and shivering, and had a bronchial infection, which he had had for some time. With Mrs. Hastings' care Kivu improved greatly, put on weight, and doubled his value. His original captor, according to Aspinall, was Heinz Demmer, an ex-Wehrmacht officer and now a hunter and also an animal dealer. He had been the dealer who had negotiated the sale of Chi Chi, the panda, to the Regent's Park Zoo in London. He obtained Kivu on the slopes of the Mountains of the Moon, in East Africa, regretfully by shooting the baby's mother.

Mrs. Hastings has described some of the problems and pleasures of bringing up a baby mountain gorilla. She said that he eats as much as a child of four. About 6:00 A.M. he is called and given two glasses of milk, and at 8:15 he has a human breakfast of cereal with sugar and milk. At 11:30 he has his orange juice and sits at the table with the family for lunch. He is very fond of curried rice, and he likes salad with French dressing. He goes to bed when he feels like it and tells Mrs. Hastings when the time has come by taking her by the hand and leading her to the bedroom. Every morning he has a bath, during which he is soaped over with a Chanel shampoo and then is rubbed all over with Helena Rubenstein skin food. Apparently he enjoys this very much. Mrs. Hastings told Dominick Elwes, a writer visiting the estate to see the gorillas, that "this is not so silly as it sounds. His skin is apt to get dry and scruffy."

According to Mrs. Hastings, Kivu could understand a limited number of words. He could be asked to fetch a toothbrush or a pot when they were requested by name, but sometimes he didn't feel like doing it. He was taught to brush both his teeth and his hair, though he did them rather clumsily. He would also present his hands when asked to so that they could be washed. He was a very clean animal, he had good eating habits, and his manners at the table were very good. He also enjoyed chewing gum. He didn't like to learn fancy tricks, but would, however, play a variety of games, for instance, hide and seek with children. He was also found to have developed the ability to kick a football, and he watched television, enjoyed music, and seemed to like Bach chorales best. He also liked alcohol, particularly lager beer, which he would steal if he saw a bottle around. He was also fond of sherry, claret, and gin and tonic, but he did not like whiskey or brandy. He was very strong, with very well-defined likes and dislikes. He liked women, but disliked men, and was inclined to give men a sharp nip on the forearm. He had tremendous affection for Mrs. Hastings. When she was asked if she would go into his cage when he was full grown, she said, "Would I go into his cage when he is full grown? But of course. I'm his mother."

Lillian Russell Rigby wrote a book called *My Monkey Friends,* and in it she has interesting accounts of some of the pet gorillas that she met. The first one she actually came in contact with was about eighteen months old, and he was a visitor for a few weeks at Dr. Albert Schweitzer's hospital in Lambaréné. She found the animal "a very gentle, wise, dignified, attractive little person." She said that she could not look at him without getting the impression of sadness. Although he was a baby, she felt all the sorrow of the world seemed to be in his deep brown eyes, which looked so humanlike. She said it is no wonder that he had this sad look since his mother had been murdered by a hunter, and there was no one else that the baby could turn to to take a mother's place at that time. The man who owned him had planned to take him to Europe and sell him to a zoo. The baby's future was bleak; it meant probably tiresome months in some kind of menagerie and then early death, which was the fate of most gorillas in captivity during the early part of the century. He left the hospital finally, and they never heard of him again. Eventually the French government forbade the removal of anthropoid apes from its equatorial provinces. There is still a limit on the export of apes from these countries even though they are now under the control of their own people.

The second gorilla that Ms. Rigby came in contact with was called G'rillali, and he was also headed for Europe and probably a menagerie. He was two years old when he came to the hospital for treatment as a patient. When his mother had been killed by a hunter, the baby had been shot through the hand. Everyone in the Gabon territory knew that the staff at Dr. Schweitzer's hospital was prepared to ameliorate the sufferings of animals, as well as the sufferings of human beings. G'rillali's wound healed very quickly and he was invited to remain at the hospital for several months until he was due to leave for France. In the atmosphere of the hospital he gained a great deal of affection and soon became a much happier little animal. The hospital carpenter made him a tiny table and a chair to match. He would sit at it to have his meals and permitted the staff to put a little bib around his neck without protest; he would then sit at the table and eat his porridge from a soup plate with a spoon, without any mess. In fact, his table manners were a great deal better than those of many young children. Unfortunately the time came when G'rillali had to leave, and he set off in a motorboat for the coast, where he would catch a ship for France.

Another pet gorilla acquaintance was Nestor. His role was that of a beloved only child of a Swiss merchant and his wife, who lived at Port-Gentil, which was the main port of this colony. He had very little hair. He had a little bit of the pathetic melancholy that the other two gorillas had and was an animal of little charm at that time. In fact, he behaved as if he were spoiled. He was also very naughty. He chased his new mother all over the house and also around the compound and would cling to her leg at the level of her ankle while she walked. If he was not permitted to do that or if he was left on his own, he would scream with rage. When Ms. Rigby met him a year later, however, he was quite a different person. He had grown considerably and had a beautiful coat of rich black hair. His expression had become much more serious. About five months later she saw the gorilla again, and by that time he was probably three or four years of age. He had grown even bigger in size and was hairier and very well behaved and agreeable. He spent most of his day romping around the compound with black children. Although his arms were as thick as a laborer's and he was vigorous with them, he was not unduly rough, as if he understood his tremendous strength. He was not so gentle with young trees, however, and he would often climb them and bend them until they snapped.

At meal times he was very well behaved. He was given a cushion a little way away from the table and would sit on it in a decorous fashion and eat and drink quietly and cleanly. If he wanted a second helping, he would get up and come to the side of his mistress and hold his plate out to her. With each meal he was served a wine bottle filled with milk, and he drank this very carefully, never spilling it, and if he didn't want to drink it all at one time, he would put the bottle carefully down by his side. During the evening hour, just prior to dinner, Nestor decided that this was his time to play and mingle with the grown-ups, just as a child would. Sometimes he would sit at a table, playing with a toy that he had selected from among his collection. If somebody played a bright tune on the gramophone, he would do a little jig to it. Like all small boys, he was very fond of wrestling and rough and tumble, and he was prepared to engage anybody who was willing. He particularly liked being tickled. After dinner Nestor would be taken to a big cage in the compound and put to bed in a wooden bedstead with blankets and a pillow. Each night a bottle of milk was placed by his bed so that he would find it waiting for him when he awakened in the morning. As Ms. Rigby says, "No wonder he was such a sturdy, intelligent, placid, good-natured youngster!" Ms. Rigby had one other gorilla friend. This was Seppi, who had been presented to Dr. Schweitzer in 1931. When she first met him at the Schweitzer hospital, he was a terrified little fellow, clinging to everybody that would accept him, but he soon adapted to the life at the hospital. One of the important new things they tried was to persuade him to accept more than one person as a mother figure, to get him to feel that all the hospital nurses could function in the role of mother and thus whatever happened to any of them, he would always have someone around who could give him the love and affection he desired. He soon showed that this idea met with his enthusiastic approval. Another thing in his favor was that he was brought up with two little black children who were orphans, a boy and a girl, and they were about the same size as he was. Thus he always had their companionship and was saved from the boredom of having no one to play with.

He was happiest with the boy, Albert, who was one of Dr. Schweitzer's many godsons. It is interesting that when he felt that Albert was being scolded without proper justification, he would protest and try to defend him. The three were taken for a walk each day by a nurse, and it was always Seppi who was the leader of the little procession. Always the little girl was in the rear. They all walked like gorillas, on all fours. Each was trained in personal hygiene and good manners, but Seppi was fortunate in that he was not obliged to wear clothes as the two little children were. Seppi was interested in some blind puppies. He would like to pick one up and smell it. Apparently there was something about the smell which he did not approve of because he would invariably replace the puppy in its box. If he picked up a kitten of about the same age, however, he would also smell it, but in this case would carry the kitten around in his arms for a time before putting it back. This is interesting when one recalls how Mrs. Hoyt's gorilla adopted a cat and carried it around with her at all times. The people at the hospital had hoped that Seppi would have many happy years ahead of him, although they were sorry that he was an orphan and was never going to be able to enjoy the family life that he could have had in the wild. They were also a little worried about what would happen when he became an adult gorilla with the immense strength that these animals have. The problem solved itself, however, in a very unfortunate way, when the poor creature died an early death of dysentery.

Heinrich Oberjohann, who was an animal collector, acquired a baby gorilla in Togoland, West Africa, whom he called Nyanya. Nyanya had been brought up with people and had never lived with gorillas. The only company he had ever had was an old African who acted as his adopted father. Once when staying at an encampment with other blacks, an interesting incident occurred. The gorilla had been trained by his African owner to do early-morning chores such as collecting water. In the morning at this encampment, Nyanya had got up and searched for buckets to get water as he had been used to doing. He picked up two of the cook's pails, but the cook saw him making off with them and tried to stop him and get them back. When the gorilla would not give them up, the cook took a stick and threatened the animal with it. Nyanya, unaccustomed to being threatened and not liking it very much, seized the cook and threw him down on his face, raked his back with his nails, and did a war dance on his back. Later peace was obtained by permitting the gorilla to do all the morning chores he wanted to do, although there were complaints from the cook, who said that when the ape learned to cook, he would get fired.

Nyanya was fascinated by the sight of his master taking a bath and made many efforts, by rubbing himself, to produce the lather that his owner produced. Given a piece of soap to do the same thing, his first act was to eat it in one gulp. Given a second piece of soap and shown how to use it, he was able to produce soapy lather on his chest to his heart's content. After having done this, he still swallowed the remains of the soap. Nyanya caused concern on another occasion when he chased the laundry boy out and took over the laundry himself. He scrubbed soap right through the clothes, wrung them in a way they had never been wrung before. When he found himself enveloped in soap bubbles, he stormed at them and beat them off with his arms, tried to catch them, and was delighted when they burst between his fingers. Finally, becoming bored with everything, he picked up the table, the basins, the tubs, and the laundry and threw them in all directions.

Nyanya's posture was surprisingly erect, like a man's, and not

Nyanya gets a bath from Heinrich Oberjohann. From Heinrich Oberjohann, *My Friend the Chimpanzee* (translated by Monica Brookshank; London: Robert Hale, 1957).

at all like the half-dropped position, walking on the knuckles of the hand, which is so characteristic of gorillas.

Nyanya wanted a share of everything. When he saw that his master had a watch, he wanted a watch, so he would take his master's watch off his arm or out of his pocket. When the master received a present of cigars, Nyanya had to have a cigar too. So eager was he to have one of these strange objects that he stole one and climbed up a mango tree to smoke it. To teach him a lesson, his owner drilled a hole lengthwise through one cigar, filled the cavity with gunpowder, and closed it with a paper pellet. Nothing happened for two days, but on the afternoon of the third day he heard an explosion followed by a howl, and Nyanya appeared with his hair standing on end looking very disturbed. The next time he was offered a cigar, he was sitting at a table. When he saw it, he let out a scream, jumped up, swept everything off the table, and ran howling out of the tent. His interest in cigars obviously ceased from then on. This animal must have learned not only to smoke from watching Oberjohann do it but also how to use matches in order to light the cigars. This represents a considerable degree of intelligence and also manual dexterity. Although a number of the chimpanzees at Yerkes have learned from time to time to smoke cigars or cigarettes, we have not yet seen any one of them actually strike a match and light it, although we have seen them hold a cigarette to a flaming match in order to get a light.

Nyanya was fascinated by anybody wearing a skirt. He particularly liked women and girls, and they could do what they liked with him. He was not even concerned if they smacked him for being naughty. Whenever he saw a woman in a skirt, he would run up to her and attempt to embrace her. With the African women this never occurred because they were terrified of him, and if he would approach them, they would turn and run screaming away out of his reach. On one occasion a priest who was a missionary came to the camp. The moment the priest came into sight Nyanya thought from his long skirt that he was a woman and took off toward him. Despite calls from his owner to come back, he continued to rush to the priest. As soon as the latter saw the gorilla rushing at him, he went into an immediate panic. He picked up the hem of his cassock and ran like a deer. Nyanya, however, was a faster runner and very soon caught up with the priest and flung his arms around him and clasped him to his breast. At this the priest thought the end had come, and he fainted away. Nyanya took one look at the face with hair on it, since the priest had a beard, and was so shaken by what had happened that he dropped the priest and returned, growling, to the camp. After that incident Nyanya never again made a demonstration toward women wearing skirts, so it must have been an event that was beyond his comprehension.

After his African foster father died, the gorilla became more and more attached to Oberjohann and, in fact, shared the same bed with him and at night slept with his arm around him. He went everywhere with him in the day, even taking part in the major trapping expeditions which were organized to catch wild animals. On one occasion when he was not going to be taken on a long journey with the African bearers, he got very angry. Eventually he followed them and insisted on having something to carry just as the bearers did. He always carried his loads in a very awkward fashion, and after a mile or so he would always want to rest. Once he struggled on after the caravan for a considerable time, and then he suddenly found a practical method of being able to carry the load and still walk without difficulty. He then was very quickly up with the leaders, and the moment they halted for the night he would even help the cook collect firewood, fetch water, pick wild fruits. On one occasion the bearers faltered and came to a halt because Oberjohann had become annoyed when one of them stole his personal possessions. They finally finished up by attacking him, and he was saved by Nyanya, who grabbed one of them by the arm and threw him against the tree trunk, followed by another. With this the other bearers scattered like birds.

It was obvious that it would not be possible to leave Nyanya in Africa when Oberjohann went back to Europe since Nyanya was now so dedicated to him. Since he had to go back to Europe with his load of animals, the only alternative was to build a suitable cage that would house Nyanya. This the gorilla, needless to say, thought of only as a prison, and he was very resentful at any attempt to get him into it and appeared to feel that this was a betrayal by his master. Eventually, however, he adapted to the cage and joined the cavalcade to the coast. He was loaded with all the other animals on a ship for the journey to Europe. At first Nyanya behaved himself, but once he got out of his cage and found Oberjohann's collection of poisonous and nonpoisonous snakes, which he was bringing home to sell to European zoos. He picked up the crates full of snakes and threw them into the sea. When the entourage disembarked in Europe, Oberjohann had no place to put Nyanya while he carried out his business activities, so he left him with a friend, who felt it unkind to keep an animal such as Nyanya in a cage. As a result, of course, Nyanya caused all kinds of trouble, and he eventually had to be taken by his master in a taxi to be left with someone else. In the taxi he was attracted by the driver's hands manipulating the steering wheel and finally grabbed the wheel. The driver wrenched the wheel straight to prevent an accident, skidded on some tram lines, and stalled the car at a right angle across the street, blocking traffic from both directions. At this point Nyanya climbed over the seat, alongside the driver. When the driver refused to let him fool any longer with the wheel, he opened the door of the taxi and threw the driver out. He then sat in the driver's seat, pulling all the knobs and levers, pressing buttons, manipulating the gears, and blowing the horn. The last attracted a policeman, who came to investigate. Oberjohann said that they had just stalled, but the policeman saw the gorilla in the driver's seat and was rather shaken by the sight and demanded to know what he was doing there. Nyanya by this time had climbed out of the driver's seat and ran over to a fruit shop, where he proceeded to demolish the window and pick up a bunch of bananas, which he carried back to the car. The policeman threatened to arrest Oberjohann, and the gorilla immediately went for him. At this point the policeman took off at high speed. In the meantime, Oberjohann gave 100 marks to the keeper of the fruit stand to take care of the window, put Nyanya in another taxi, and drove to the police station and explained the situation to them. Nyanya by this time had calmed down and won over all the policemen at the station, and he and his master were allowed to depart in peace.

Still without a place to keep his animal, Oberjohann remembered a zoo director he knew and took the animal to the zoo, where he was accommodated in a large cage. A year later he came back and found that Nyanya was being shamefully mistreated, being flogged with sticks and treated like a wild and ferocious animal. An elderly lady who was a frequent visitor of the zoo became very fond of Nyanya and was distressed by the treatment being meted out to him. She made contact with Oberjohann, who removed Nyanya from the zoo and let him go to live with the woman. There he lived for two years completely unrestricted and very happy, but unfortunately he died later of an abscessed jaw. In a letter to Oberjohann telling him about Nyanya, whom Oberjohann described as his adopted son, the old lady wrote, "I have had Nyanya buried like a man.

He deserved that. He lived and died like one among us." Nyanya was a real ape person.

Recently in West Africa a French family, the DeMeyers, who lived in Gabon, decided to adopt a baby gorilla and, having enjoyed the experience, adopted others until they now have nine. It all started four years ago when some Africans brought them an orphan gorilla whose mother and father had been speared. Later they brought other orphan gorillas. Their weights, medical histories, and other details have been recorded. Gorilla babies, like human babies, are virtually helpless when they are first born. Like the chimpanzee, they grow much faster physically than the human baby. Mentally they grow at about the same rate as the human up to about a year or eighteen months.

Arthur, one of the gorillas in the group, is now age four and has to be caged during the day; otherwise he would tear the place to pieces. Another baby, three years old, Zozotte, specializes in scaring Africans who visit the home because he likes to see them run. Another of the three-year-olds likes to chew on a rug before it drops off to sleep. A circus once offered Henri DeMeyers $16,000 for Arthur, but was turned down flat. Henri's answer was, "We do not sell our children."

5

"What's a Nice Gorilla Like You Doing in a Place Like This?"

THE public's curiosity and fascination for gorillas which began in Europe in the middle of the last century created a demand for this rare animal as an attraction for traveling menageries and zoological gardens. Beginning with Jenny, the first gorilla to reach Europe, in 1855, who died soon after arrival, the story of the early gorillas in captivity is a tragic one. Of the approximately twenty gorillas that were imported into Europe and the United States before World War I, only two lived beyond a year. Most of them died after a few weeks in captivity. Naturally the zoos and menageries made every effort to keep their valuable exhibits alive, but little was known of the gorillas' habits and diet, and they succumbed usually to respiratory diseases or malnutrition.

It was not until later in this century that primates came to be recognized as valuable tools for research at scientific institutions and zoological gardens, and it has only been in recent years that zoos have regarded breeding gorillas in captivity to help perpetuate the species as one of their key responsibilities.

As mentioned, the problems of caring for gorillas, particularly during the last century, took a heavy toll, but after World War I and through 1920 these animals began to survive longer, and their numbers in captivity began to double in Europe and the United States.

During the time between the two world wars twenty-six gorillas were exhibited in international zoos. During this period several good longevity records were noted in America and Europe. At this writing, the record is held by Massa, a male lowland gorilla at the Philadelphia Zoo, who arrived in this country in 1931 and is still living today, having thus far spent forty-four years in captivity. Two other gorillas have lived for more than thirty years; a half dozen have spent more than twenty years in captivity, including the famous Guy at the London Zoo, and at least fifty or sixty have surpassed the ten-year mark, an achievement which would have been considered impossible about forty or fifty years ago.

The tragic history of the attempt to introduce gorillas to zoos has now had a happy ending, and the zoos are emerging as major centers for the breeding and perpetuation of this species.

In 1875 a German expedition in Loango in West Africa was given a young male gorilla by African hunters. When it was first received, it was in a deplorable condition, but it was fed well on milk, vegetables, and fruits and was restored to more or less normal health. Eventually it was shipped to Germany. During the voyage the animal was chained up, but soon became tame enough to be allowed to wander freely about the ship with some supervision. It was of a gentle disposition and was well behaved in taking food. It arrived safely in Berlin, where it became an inmate of the aquarium. There it throve at first and was docile enough, though inclined to be mischievous. Eventually it succumbed to the malady which sooner or later carried off so many of the manlike apes in a northern climate; it died of rapid consumption in the autumn of 1877 after having lived for fifteen months in Berlin.

With the help of Dr. Eduard Peuchuel-Loesche and Dr. Julius Falkenstein, a second gorilla was obtained from the Loango district and safely transported to Berlin, where it arrived in the early part of 1883 and became known as the Falkenstein gorilla. It lived only fourteen months in the aquarium and also died of TB in the spring of 1884. There was another gorilla in Berlin in 1881, but it died soon after arrival.

The first gorilla to arrive at the Regent's Park Zoo in London made its debut in 1887 and created a sensation in town. It was about three years old and stood 2 feet 6 inches in height. It arrived in miserable condition and refused to leave its traveling cage, and it had to be forced into its permanent home. A young macaque monkey was assigned to it for company, but the ape was too indisposed to return its friendship. Skillful attention and a diet of bread, milk, and fruit kept it alive for two months, and then it died. In the meantime, it had made friends with its keeper and showed signs of considerable intelligence.

The second venture was made in March, 1896, when the largest gorilla ever imported into England was obtained. It was just acquiring its permanent teeth, but though for a time it throve, it never showed much activity and died within five months. Two young female gorillas were secured in 1904 and did not survive a month, dysentery being responsible for the death of both. Still persevering, the Zoological Society of London secured in the following year a female gorilla of tender age which was known for a brief spell as Miss Crowther. She had to be forced from the cage in which she had traveled into the cage adjoining the chimpanzees. Her arrival caused the wildest excitement among the latter, who shrieked and shook their cage and danced with what has been described as "rare vivacity." The stranger gave much trouble to her keeper, seeming determined to starve, and while life remained in her, she was transported to New York. The sixth gorilla arrived in March, 1906. She was a young female and was so weak and fragile that she was never exhibited in public, being kept in a cage in a secluded part of the London Zoological Gardens, whence she would be carried in the arms of a solicitous keeper for daily airings in the sun. From the first her case was hopeless.

A physician attended her daily. Her diet and medicine were the subject of anxious consultations among the learned of the medical profession. As a leading official of the gardens remarked, the young gorilla could not have been more anxiously watched had she been a millionaire's daughter. Pneumonia caused her death, and after all their experiences the authorities reached the conclusion that it was almost an impossible task to rear a gorilla in their gardens. It is now agreed that unless gorillas can be caught young and kept in captivity near their home for some time before being brought away, thus giving them a chance to become accustomed to hand feeding and confinement, it is hard to rear them in captivity. Gorillas are now listed as an endangered species and, fortunately, can no longer be imported into many countries and, in theory, at any rate, cannot be exported from their country of origin. So at least the number of youngsters that have to suffer the traumatic experience of being brought from their homes in captivity is substantially reduced.

When a gorilla has made a successful transition to the captive life, he can show signs of being very happy.

An issue of the London *Times* in the early 1900's describes a young gorilla in captivity.

I found the creature romping and rolling in full liberty about the private drawing room, and how looking out of the windows was all becoming gravity and sedateness as though thoroughly interested in, but not disconcerted by, the busy multitude and novelty without; and bounding rapidly along on knuckles and feet to examine and poke fun at some newcomer and playfully mumbling at his calves, pulling his beard—a special delight—clinging to his arms, examining his hat, not at all to its improvement, or curiously inquisitive after his umbrella, and so on with visitor after visitor. If he became overexcited by the fun, a gentle box on the ear will bring him to order like a child—like a child, only to be on the romping intermediately. He points with his index finger, claps his hands, puts out his tongue, feeds on a mixed diet, decidedly prefers raw, roast meat to boiled, eats strawberries with delicate appreciativeness, is exquisitely clean.

In 1883 France exhibited its first gorilla, a three-year-old male. It was scientifically described by Alphonse Milne-Edwards in 1884. This animal survived for only a short time, and it was said to be savage, morose, and brutal.

There had been a gorilla in Germany for some time, in the early part of the twentieth century, which was shown in the zoo at Breslau. She was called Pussi and lived for about seven years. Pussi had actually been purchased in Liverpool in 1897 when she was four years old and weighed thirty and a half pounds. She died in October, 1904.

In 1906 Dublin obtained a young male gorilla for exhibition, but he lasted only a few weeks. Subsequently in January, 1914, a young female gorilla arrived in Dublin from the Gabon. This animal was called Empress and it lasted for three years and four months.

Sometime in 1898 the first gorilla to land alive in the United States arrived in Boston, but the poor animal lived only a few days. It was never exhibited in any zoo or even in a circus.

Richard L. Garner brought a female gorilla, two and a half years old, to the Zoological Park in New York in 1911. It had come from West Africa and lasted only about a month, arriving on September 23, 1911, and dying on October 5; according to the record, it died from malnutrition and starvation. It weighed only twenty-five pounds.

The Zoological Park received a second gorilla on August 21, 1914. It was called Dinah and survived for nearly a year, eleven months and ten days, in fact. The animal had been held in captivity for two years. For one year of this time she had been cared for by Garner, who was acting as the agent for the New York Zoological Society at that time. The animal was alleged to have died of malnutrition and rickets. She weighed forty and a half pounds on November 15 and was three and a half feet high. She was probably about four years old when she died. This gorilla, Dinah, was one of two which Garner had kept in captivity in Africa. The other one was Don, and he lasted only three months after he was captured and actually died in Africa. Dinah was the last gorilla to come out of Africa before the First World War broke out.

Dinah landed in New York on August 21, 1914. She did not object to civilized foods at all and ate all kinds. She was put on exhibition in the primate house, and many people thronged the house to see her. Her keeper was Fred Engeholm, and she became very attached to him. She also made friends with quite a lot of other people. Her list of friends included secretaries, directors, curators, keepers, reporters, and photographers. She had no objection to publicity and cheerfully posed for both moving and still pictures and made every effort to fill the high position which she obviously occupied in the world of zoology. She had a very black shiny face which was like polished ebony and also very large liquid brown eyes.

Garner said that Dinah was the only gorilla he had ever seen that attempted to laugh or even to smile when he tickled her under the arms or on the bottom of her feet; when he did so, she would chuckle almost to the point of laughing. She would constantly challenge him for a romp around, and when he entered her cage, she would climb on his shoulders or his head and slap his cheeks, hit him hard on the back, or snatch off his hat. He said that she had a real sense of humor that manifested itself in all kinds of pranks. She had a trapeze in her cage, and she used to pose on that in a unique fashion. She also liked to play solitaire football and she would clutch straw in her feet and, using her arms as crutches, rush across the floor and toss the straw against the wall. She would then catch it in her hands and play with it. Occasionally she would rise to an erect position and beat her chest with her hands. Garner gave her a mirror, and she actively searched for the animal behind the mirror as most animals do when first presented with something of this sort. She would peep over, under, and around the sides of the mirror. In the process of doing this, she expressed interest, anxiety, and disappointment, with a humanlike expression.

Dinah eventually showed some neurological symptoms. Her appetite fell off, and the muscles in her arms and legs did not work well. She looked as though she had a case either of loco-motor ataxia or infantile paralysis. She was examined by scientists of the Rockefeller Institute for Medical Research, but did not survive.

The first mountain gorilla to come to the United States arrived in New York toward the end of 1925; her name was Congo. The gorilla was one that had been captured by an American hunter, Ben Burbridge.

Between 1927 and 1928 five other young gorillas arrived in the United States from West Africa. One of them was purchased for the Philadelphia Zoological Garden. His name was Bamboo, and he died at the ripe old age of thirty-five, apparently of some heart disease. Prior to that he had shown many symptoms of growing old.

In 1933 the London Zoo bought a pair of gorillas, for £2,000, which had come from West Africa. They were seven and five years old when they were obtained, and they weighed 112 and 70 pounds respectively. Their names were Moina and Mok. Mok's head at the time he was bought was round like a chimpanzee's. Three years later he developed a big crest on the back of his head, and Moina a much smaller crest. Since those early days zoos all over the world have obtained gorillas for exhibition. In 1960 there were 128 gorillas in captivity in different parts of the world. Of these, 87 were in captivity in the United States, 34 in European zoos, and 3 in the Ueno Park in Tokyo. There were 3 in the Nagoya Zoo and there was 1 in one of the Australian zoos.

In 1970, ten years later, the number of gorillas in captivity had risen from 128 to 134; of these, 15 (6 males and 9 females) were alleged to be mountain gorillas, although Colin Groves does not agree that this is so. Indeed, he says that they are lowland gorillas that come from the eastern part of the Congo.

The Atlanta City Zoo has a gorilla that has achieved considerable local, if not national, notoriety. Called Willie B., he is named after a former distinguished mayor of Atlanta, William B. Hartsfield. He is about sixteen years old and arrived at the zoo on May 29, 1961, from Kansas and weighed sixty-five pounds. He is now a whopping 470 pounds and looks it. The news media in Atlanta have made Willie B. a well-known personality.

Once the zoos had mastered the business of successfully

keeping gorillas in captivity, there was competition to be the first to breed them. There were many disappointments, and it is difficult to say exactly what were the reasons for this delay. There may have been a nutritional factor involved; but there must be more to it than just that because the Yerkes Primate Center, which has been very successful in breeding orangutans and chimpanzees, has had difficulty in breeding gorillas. Two gorillas were born at the center, and both are living, but we are surprised that we have not had more. Obviously there is much involved in successful breeding. Some experts say that gorillas breed best when they are kept in groups. The practice of keeping them in cages alongside each other and bringing them together for mating does not appear to be very successful. Other people who have worked with gorillas say that it is better to have a group of gorillas grow up together as adults and then breeding will take place.

To the Columbus Zoo in Ohio belongs the credit of having produced the first captive-born gorilla. The Columbus Zoo bought two gorillas, Baron and Christina, in 1951 as youngsters of about three years of age. They produced their first baby on December 22, 1956. There have been babies since, to other mothers, and a three-pound, nine-ounce female was born on February 1, 1968. The mother, Kolo, was eleven years and one month old, and the baby stayed with her only twenty-two hours before she was taken away and put in an incubator to be raised by hand.

Not long after the Columbus Zoo birth, the zoo in Basel, Switzerland, produced an offspring, the first gorilla born in Europe. Its mother was known as Achilla and the father was named Stephi. Goma, the product of the mating, was born on the night of September 22 or 23, 1959. Achilla had come to Paris with a shipment of animals that had journeyed from the Cameroons in West Africa. She arrived at the Basel Zoo on October 23, 1948. She weighed only seventeen and a half pounds at that time and was raised with a young chimpanzee. They played together and, according to Dr. Ernst Lang, director of the Basel Zoo, gave every appearance of having a happy childhood.

In 1952 Achilla complicated life for everybody. She was drawing with a full-color ball-point pen and something frightened her, so she put the pen in her mouth and swallowed it. A foreign object of this size cannot be permitted to remain in the stomach, but gorillas had never been operated on before; and the veterinarians were very concerned about the anesthetic. The animal was too strong to be held down, so the vets first gave her a drink spiked with a soporific drug. This made her react drunkenly. A half hour afterward she was still conscious. By this time a pediatrician, a children's surgeon, and a physician from the Basel Hospital had arrived; in addition there were the zoo veterinarians and Dr. Lang himself. Dr. Lang then gave Achilla an injection of morphine that would, as he said, "send a man straight to dreamland." They had to give her a number of shots before they were able to make her somnolent and get her on the operating table. Once there, she was attached to an anesthetic machine. It was not long before the stomach was opened and the ball-point pen removed. When they finally got her back in her cage, however, she did not wake up. Her pulse had slowed, and her respiration was scarcely detectable. Dr. Lang almost gave up hope; but finally she was given a heart stimulant and heavy massage, and she opened her eyes. From then on, she recovered fully and the wound healed satisfactorily. When the scar healed, she removed the stitches herself without bothering to have the doctor do it.

Willie B., Atlanta Zoo's famous gorilla.

100

Gorilla Achilla nurses her baby, Goma, at the Basel Zoo. From E. M. Lang, R. Shenkel, and E. Siegrist, *Mutter und Kind* (Basel: Basilius Press, 1965).

Kathyrn and baby Frederika. (Courtesy of Jim Ellis and Oklahoma City Zoo.)

In 1953 the zoo in Basel was able to get a seventy-four-pound male gorilla from the Columbus Zoo. They put Achilla and the recently arrived male, Stephi, in neighboring cages. They were able to smell each other through the bars, but were suspicious of each other in the beginning. After a time they began to put their fingers through and play with each other. Eventually they were well enough acquainted that they could be put together in the same cage. Dr. Lang was thrilled because the Basel Zoo was at that time the only zoo in Europe that had a breeding pair of gorillas, and the zoological profession in Europe was eager to see if breeding would be successful. Achilla had reached 154 pounds in weight and was about four feet in height. Stephi had reached 310 pounds and was over five feet. By March, 1959, they found that Achilla could not tolerate the male in the same cage except for a very brief time. As she got bigger and bigger, it was obvious that there was a baby on the way. On the night of September 22–23, 1959, Goma was born. The zoo staff did not see the birth, but when the keeper came to her cage early in the morning, Achilla already had the tiny gorilla baby in her arm. Its face was pale with a rosy tint, and it had a good cap of black hair. The mother was not holding the baby to her breast, but had it on her left arm with the face turned so that it was away from the breast, and the baby was not able to seek for milk itself. After thirty-six hours she had still not nursed it, so they decided to give Achilla a little sedative. When it took effect, she put the baby down and went off to lie down. Then they were able to go in and get Goma. Goma was then taken over to the Lang household to be brought up by them. She demanded food about every two hours and drank up to 33 cc (a little over an ounce) at a time.

Goma's birth was a great personal triumph for Dr. Lang and for the Basel Zoo. She was the focus of reporters, photographers, and television cameramen. Books written about her were translated into a number of languages, and a number of scientific papers were written as well.

Now Goma herself has a baby, so that the Zoo in Frankfurt has two generations of captive-born gorillas.

The details of Goma's upbringing make fascinating reading and were published in a book called *Goma, the Gorilla Baby,* written by Ernst Lang and profusely illustrated with photographs.

Achilla, on April 17, 1961, produced a beautiful, healthy male gorilla which she mothered in a very normal fashion. Because of this, the baby, instead of being taken away from her as was the first, was allowed to remain with her. When the keeper entered the gorilla house a little after seven one morning, there didn't seem to be anything particularly unusual about Achilla. She was simply sitting in her rubber motor tire, and she grunted what Ernst Lang described as a "good morning." At about half past seven the keeper had planned to let her roam about in her outdoor cage. Before he could do so, however, she got up, reached underneath her body, and produced a baby gorilla. She had given birth to the gorilla apparently in a standing-up position. She then came to the front of the cage and sat in the rubber tire, and the umbilical cord, to which the baby was still attached, was still hanging from her. This unusual sight was noticed by two chimpanzees who were in the next cage, and they began screaming with excitement. In the meantime Achilla began to build herself a nest in her rubber motor tire. She got hold of a jute sack and spread it in the tire and lay in it. She held the baby in her left hand most of the time or else she put it in her arm and pressed it against her body. Later in the day Achilla showed her infant through the wire netting to the chimpanzees in the next-door cage and also presented it to a lady that she knew and was fond of. At a later stage she was observed lying back on her tire with the baby in her lap. The baby was waving its hands and seemed to be very contented. On the second day, as Achilla was lying on her back with the baby lying on her belly, it suddenly began searching round on the chest and particularly in the region of the armpit. Finally it located the mother's teat and immediately began to suckle. Later on he searched for the breast again and was successful. The mother did not appear to play any part in putting the infant up to the breast to suckle. This was the second day, and on the third day the baby gorilla and his mother were permitted to go out onto the open-air terrace, where the mother was able to sit in the sun. She wandered around the enclosure from time to time and carried the infant in many different positions. It took a few days for the baby, Jambo, to learn to hold on properly to his mother with his hands and feet. The head keeper came back from a holiday after Jambo was eight days old, and Achilla was very pleased to see him and grunted happily and held the infant toward him behind the wire netting. When he went into the cage with her, she reacted in a peculiar way by going into a corner and turning her back to him. By the afternoon, however, she was agreeable to sitting down with him at the table they had used in the past, and she ate some soup provided for her with a spoon. On two occasions she tried to put the empty spoon to Jambo's mouth. When Jambo turned its head away, she tried it in Jambo's ear, which was equally unsuccessful. On the next day after his return the head keeper, Carl Stemmler, was permitted to stroke her little Jambo on the head. There seemed to be some evidence that Achilla was not producing enough milk for the baby, which was still pretty scrawny, so additional food in the form of milk and a fortified milk was added to her diet with the hope that this would stimulate the production of more milk in her mammary glands. When Jambo was fifteen days old, Achilla greeted Dr. Lang by presenting her baby to him behind the wire and actually pushed one of Jambo's hands out through the wire so that Dr. Lang could take hold of it. She seemed to be very pleased that Dr. Lang was fussing over her baby.

At two weeks Jambo was doing very well and had rounded out nicely. The keeper was able to examine and stroke him. Sometimes Achilla would carry her baby around wrapped up in a sack. She would occasionally moisten the tips of her fingers and wash Jambo with them. When Jambo was two months old, he was handed over by Achilla to the keeper, Carl Stemmler, and he was permitted to hold him for the first time. Achilla didn't seem to trust him very much, however, because she took the baby back straightaway. After that occasion whenever the keeper asked Achilla for the baby, she would hand it over and in fact behaved as if she were glad to have an occasional break from being a mother. The baby itself didn't seem to mind being handed around, nor did he seem to worry how his mother carried him—whether it was on her leg or under her belly or in her hand.

At two months Jambo began to get additional goodies, such as a little orange juice and even a few slices of banana. He also got a few drops of multivitamin preparation each day. By about twelve weeks Jambo showed very clearly that he wanted to be erect and would pull himself into the sitting position on every occasion—even after his mother had carefully laid him down on his back.

Frankfurt has also been very successful in breeding baby gorillas. It has had five now, four of which were offspring of a single female, Makula. The fifth animal was called Dorle, and his birth was very fast, taking only a few seconds. The mother had been seen playing with a group in an outside enclosure, and the birth took place there, the mother seeming to be as surprised as the keeper when the baby arrived. Although she looked at the baby, which was shouting and whimpering, she did not touch it or pick it up. So the keepers had to go into the compound and retrieve it; they took it immediately to the nursery. The animal weighed two kilograms, about five pounds, and he was bottle fed. Two gorillas have been born in zoos in Japan. The first was born in the Kyoto Zoo on October 29, 1970, and on February 2, 1971, in the Ritsurin Zoo in Takamatsu, another gorilla was born.

The Basel Zoo has now had six births. Jambo was born in April, 1961, after the mother had had a 252-day pregnancy. The mother again was Achilla, so Jambo provided a brother for Goma, and there is also another brother who has been born.

The firstborn gorilla at the National Zoo in Washington was born on September 9, 1961, shortly after 6:00 A.M., and was the second gorilla to be born in an American zoo. The baby's name was Tomoka, and the mother and father were Nikumbo and Moka. The same pair produced Leonard on January 10, 1964, named after the late Dr. Leonard Carmichael, the secretary of the Smithsonian Institution and chairman of the Yerkes Center's Board of Advisers for many years. Leonard went to the Toronto Zoo and unfortunately died there.

On April 8, 1967, the same productive pair of gorillas produced a third offspring—a female named Inaki. Inaki came to the Yerkes Center when she was a year old; she became one of our best-loved gorillas and did a fine job helping Dr. Russell Tuttle in his study of the electrical changes in the arm muscles during her knuckle-walking and hanging by her arms.

The Cincinnati Zoo has two breeding pairs of gorillas, King Tut and Penelope and Hatori and Mahari. Each of these pairs has produced four offspring, making the Cincinnati Zoo the holder of a world record for eight gorillas born at that zoo.

In 1973 the Jersey Wildlife Preservation Trust, which maintains a small select zoo, had two baby gorillas born during the summer. They were called Assaumbo and Mamfe. Dr. F. Stephen Carter, who is a consultant pediatrician to Jersey (an island between France and England), made a study of these little gorillas and compared them with humans. As far as their neuromuscular performance was concerned, he said that the human infant is much more immature at the time of birth than is the baby gorilla. The average length of gestation for the gorilla is about ten days less than it is for humans, and humans would require a longer pregnancy to produce the same degree of maturity at birth that the young gorilla has. If the human gestation period were longer, however, the human infant would probably develop a head that would be too large to come down the birth canal.

The skull bones of the human baby appear to be capable of much more movement than the comparable bones of the gorilla, and they appear to be much softer. The bones of the human skull may even overlap each other in the process of being born. The bones of the gorilla skull, on the other hand, are very hard and relatively immobile, so that it is unlikely that any molding of the skull could happen during the birth process. Human babies have two fundamental reflexes which are probably of little value to them at their present stage of evolution, but which have been carried by them through a period of many millions of years of evolutionary development. These primitive reflexes were probably important to them from the point of view of survival in their early evolutionary history. One of these reflexes is called the grasp reflex. If, for example, you put a finger in the palm of a baby's hand, it will hold onto it firmly. It is then possible to lift the infant into a position in which he can support his entire weight by holding on with that hand. The same grasp is, of course, present in baby gorillas, but it is very much stronger than it is in the human, and there seems to be not the slightest danger of the gorilla relinquishing his grip.

There is another reflex called the moro reflex, also called the startle reflex. In this, if you let a baby feel that he is about to fall or is falling, he will immediately throw his arms and legs forward and fan out his toes and fingers to enable him to clutch anything, for example, the hair of the mother, if he were a gorilla. Needless to say, the same reflex is well developed in the baby gorilla, but the speed at which the reaction occurs is much faster than that in the human infant. Both these reflexes have no practical value to the human baby and make sense only if they are thought of in terms of the evolutionary development of humans.

The psychological development of the baby gorilla is also much ahead of that of the human. Jennifer Hughes and Margaret Redshore of the Department of Psychology of University College, London, studied the two gorillas at Jersey and found that the baby gorilla after two weeks would track an object visually as it was moved slowly in a 180° arc. This tracking, however, was jerky, and the animal did not maintain its tracking for the entire arc. When the object disappeared behind the infant, the gorilla's gaze came back to the midline, and he did not search around the point where the object disappeared. He could do that at two weeks; the human couldn't. At six weeks the gorilla followed an object smoothly and continuously as it was passed through a 180° visual arc. His gaze lingered at the point where the object disappeared. The human could not do this at six weeks. At eighteen weeks the gorilla could obtain a completely hidden object by removing the cloth and grasping the object. The human baby was not able to do this for twenty-nine weeks. In addition to these psychological developments, in the area of motor activity the gorillas were very much ahead of human infants. They reached for and would grasp objects that they could see in their visual field before a period of two and a half months. A human infant does not show this behavior until sometime after four months.

Baby gorillas weigh 3½ to 5½ pounds at birth, which is a good deal less than most human babies, although they grow to a much greater weight than that of the human. The rate at which they accumulate weight is faster than humans, although there is a good deal of variation among the individuals. Some gorillas, for instance, will double their birth weight at nine weeks and some at seventeen weeks, and then there is another doubling of the weight between twenty-five to forty weeks. Up to the fifth year the weight of both males and females is pretty much the same, but from six years on there is a rapid difference. For instance, the average weight is as follows: At one year a gorilla weighs 15 pounds, at two years 30 pounds, at three years 55–60 pounds, at four years 70–80 pounds, at five years 110–120 pounds. The female at six years is 125–140, at seven and a half 160–175, at eight years 200–210, and from then on they stay pretty much in that range with a further weight increase later. In the case of the male, at six years he is 145–170, at seven years 172-200, at eight years 225-250, at nine years 290-350, and at ten years 330-360.

Of course, 330–360 pounds is not the top weight that gorillas will get to. Gorilla Guy of the London Zoo reached his top weight at the age of seventeen years. He now weighs between 460 and 480 pounds. Gargantua, who belonged to Ringling Brothers and Barnum & Bailey Circus, at his top weight weighed 550 pounds. Bushman from the Lincoln Park Zoo in Chicago weighed about the same. Mrs. Hoyt's Toto was a very big female and reached 438 pounds.

Colin Groves has given a list of the various activities and

developments of the gorilla compared with chimpanzees and humans. These are as follows: raising the head while lying on the belly—in the gorilla 1 week, in the human 5 weeks. Raising the head while lying on the back: the gorilla 3 weeks, chimpanzee 5 weeks, human 6 weeks. Crawls steadily: gorilla 9 weeks, chimpanzee 15 weeks, human 37 weeks. Stands up steadily: gorilla 17 weeks, chimpanzee 21 weeks, human 41 weeks. Stands erect bipedally: gorilla 20 weeks, chimpanzee 26 weeks, human 33 weeks. Walks first step upright: gorilla 34 weeks, chimpanzee 30 weeks, human 52 weeks. Sexually matures: 6 years for the gorilla, 6 to 7 years for the chimpanzee. Our experience with gorillas, however, is that they are much older than 6 years, at least in captivity, before they become sexually mature.

On July 23, 1974, the Gladys Porter Zoo in Brownsville, Texas, had a new baby gorilla, Sukari, the offspring of Lamydoc and Katanga, who delivered her without any assistance. The mother was seen to be behaving in a rather strange manner by the night keeper, Otto Posler. She was very secretive and kept moving behind stone pillars so the keeper couldn't see her. Eventually, however, she was lured into her den by the evening meal. When she came in, it was seen that she had a newborn baby. Katanga has been three times pregnant in the last three years, and she aborted a six-month fetus, which terminated her second pregnancy. Her first baby died at seven months when it got a virulent attack of gastroenteritis.

Katanga reared her third baby very well for the first two weeks and seemed to enjoy fussing about with it. Then, for some reason or other, she became disenchanted, just as a Yerkes Center gorilla mother, Paki, did with her first baby. The night keeper found her trying to cover the baby with her rubber water tub. She was very rough with it and dragged it around by the foot. It was decided to remove the baby from her. She had to be immobilized with an anesthetic before this could be done, and the baby was finally retrieved. She weighed six pounds and two ounces and had been rather badly scraped and bruised from the mother's treatment. She was placed in an incubator, and for a time the little animal would not take any nourishment. She kept this up for about eight hours and seemed to be very tired after all she had been through. The next morning, however, she did begin to eat. She had a good deal of blood in her urine, which suggested some internal injuries as a result of the rough treatment she had received. She lost weight and had to be fed intravenously and by stomach tube.

Fortunately she responded and by the fourth day had started to eat well.

The Yerkes Center has had two gorilla births. The first of them was called Kishina, and her mother was Paki, who was also the mother of the second gorilla, born a few months ago and called Fanya. Dr. Ronald Nadler of the Yerkes Center has described the birth of Kishina. About four hours before the birth Paki became restless and agitated and could not maintain one position for very long. During this period she seemed to be straining and was obviously having labor contractions. The birth sac began to protrude from the vagina. At first Paki touched it rather gently, but eventually she tore it, and the fluid spilled onto the floor. At this stage Paki actually tore off a bit of the sac and ate it and also licked up some of the fluid. Then the head of the infant appeared in the vulva. Paki kept touching the head and her vulva and licking her fingers. Simultaneously she moved around the cage in a squatting position. Eventually she walked to one corner of her cage, lay on her side, and gave birth. The baby gave a few gasps and then began to breathe normally. At first Paki paid no attention to the baby; instead she began licking blood and fluids from her hands. After a few minutes, however, she began to handle her new baby, Kishina, and became attentive to it, licking its body and especially its head. She did not attempt to bite through the umbilical cord, but pulled it and the attached placenta until it finally parted from the umbilicus. After an hour of licking, Paki picked Kishina up and pressed her to her chest. She did this more often as time progressed. Eventually Kishina got a nipple in her mouth and, twenty-four hours after birth, began to nurse for long periods. Paki behaved like a good mother for a few days and then, for some reason, began to resent the baby. We could not risk anything happening to our firstborn gorilla, so she was taken away and put in our ape nursery, where she received expert infant care. Today she is a thriving young gorilla more than two years old.

The first gorilla from the eastern part of Africa was born in the Antwerp Zoo on June 9, 1968. The little animal was called Victoria and was the product of Kisubi and Quivu. Although they came from the eastern part of Africa, they still were lowland gorillas, not mountain gorillas. The mother was five years old when she was captured in 1963. When her baby was born, she behaved very well with it and bit the umbilical through quite soon after the birth, and within twenty-four hours she had the baby suckling. She would tend to carry the baby in the crook of her arm, and it fed very well from the breast. At three months the baby began to toddle, and the mother began to play rather roughly with it. In fact, she sometimes got so rough that the zoo staff had to remove it from her to save its life. It grew up and is in very good condition.

It is rare for gorillas, at least on the records we have, to bear more than one baby at a time, but Makula, a gorilla in the Frankfurt Zoo, produced twins on May 3, 1967. Both were females, but they were not identical twins. A female gorilla in Kansas City had previously had twins in 1974, but they both aborted. There is one record also of twins having been born in the wild. Even in chimpanzees there are only seven recorded twin births, and one case is recorded of triplets, which did not survive and which were born to a chimpanzee at the Yerkes Primate Center.

When a gorilla is newborn, he is not very black; in fact, he is only slightly grayish and is rather flesh-colored. He doesn't have much hair, although he has more on the top of the head. The hair grows very quickly, particularly on the back. The hair on the belly and the chest takes quite a long time to develop. The baby gorilla also develops a little white tuft of hair on his behind. That lasts until his fourth year, when he seems to shed it. Possibly this is some kind of visual recognition signal in the wild.

Some gorillas in zoos became famous. One of these is Guy, a gorilla from West Africa who arrived at the London Zoo on November 5, 1947. November 5 in England is Guy Fawkes Day, a day in which the English celebrate with fireworks the attempt by Guy Fawkes during the seventeenth century to blow up the houses of Parliament. So the little gorilla was called Guy because of this fact. He was about one and one half years of age and weighed twenty-three pounds. He was a very playful little creature. His keeper, Laurie Smith, would sweep his cage out every morning, and it took awhile for Guy to be trained so that he would not interfere with this procedure. If for any reason the keeper put the broom down for a minute or two, Guy would immediately assume that all the sweeping was over, and he would come running over, anxious to play. Guy never had a mate, and no other gorilla has been put in with him. It was not possible to get a mate for him while he was very young, and as an adult it would probably be impossible to get him to accept one.

110

There are many other big male gorillas that became famous, including the gorilla Alfred of Bristol Zoo. Bushman of Chicago's Lincoln Park Zoo was a national celebrity. He was captured in equatorial West Africa in 1928. He was only a few months old at the time, but Lincoln Park Zoo paid $30,500 for him. He arrived in Chicago on August 15, 1930, weighing just seventy-nine pounds. He lived to the age of twenty-one years and died on New Year's Day in 1951. Bushman was one of the finest specimens of lowland gorilla ever to be seen in a zoological garden. He was over six feet tall, which was very tall for a gorilla, and his weight was about 550 pounds. Eddie Robinson, a keeper, stayed with him from the time he came to the zoo, when he was probably one and a half to two years old, to his death.

Once he escaped from his cage because the lock had not been properly secured, but another keeper, holding a small snake in his hand and extending it toward Bushman, forced the big gorilla back into his cage. Bushman had little fear of anything else, but apparently he was frightened of snakes, and the security of his cage was the best answer to having the snake thrust at him. The Field Museum in Chicago had his skin stuffed and mounted, and he is exhibited there.

Colin Groves, writing about the gorilla, quotes a letter that was written to him about Bushman by E. Gates Priest of Des Plaines, Illinois.

Bushman had a marvelous build, over six feet tall, and enjoyed excellent health. Perhaps many gorillas could equal these qualifications, but the difference was in his character and disposition. He was friendly and a great showman too. He didn't know he wasn't a human being until he grew so big that he couldn't be allowed out of his cage and his keeper couldn't be allowed in it. Having been associated with people since his earliest days, he was very friendly with most everyone. When the weather permitted he was taken out on the park lawn to romp with his keeper and to associate with people too. He liked to play football and became quite an expert at both tackling and carrying the ball. Finally, he became so strong that when he squeezed the ball under his arm it burst. After ruining three balls in one week it seemed too expensive a pastime. Later he was given a tire from Hitler's private car to throw around in his cage.

He was intelligent, friendly, playful, a tease, sometimes jealous, and always glad to show off for the crowd. He liked most people but there were a few individuals he decidedly did not like. He tried to give these few the cold shoulder, and if they were allowed behind his cage, he would throw anything at them with unerring accuracy. I have seen him throw a boiled potato from the far side of his cage through the bars and hit a person on the forehead.

There was a chair in his cage which was attached to a scale with a dial out in front. He often sat on the chair, and would do so on command, and that showed his weight to the crowds watching him. His top weight was 565 pounds. Early on the morning of January 1, 1951, Bushman died. All the people closely associated with him mourned him as though he were a member of the family. Approximate age 23.

The Lincoln Park Zoo has another famous lowland gorilla, called Sinbad. He is the largest in its present collection, weighing 530 pounds, and he eats an average of thirty-four pounds of fresh vegetables, fruits, cereals, and whole milk each day.

In the fall of 1973 the Lincoln Park Zoo sent Kisoro, its fifteen-year-old breeding male, to Howlett's Animal Park outside London. It had four females, and the male that was with them had not been successful in breeding, so Kisoro was sent to try to help out. The Lincoln Park Zoo was just finalizing plans for its new ape house, and by the summer of 1975 it will probably have it under construction. The purpose is to have a family of apes living together. So far three baby gorillas have been born, and the zoo hopes to breed more.

The Lincoln Park Zoo had its seventh gorilla born there on September 8, 1974. The mother, Mary, was the fourth female to give birth at the zoo, and the father, Frank, was the zoo's third proven male breeder. Lincoln Park now has sixteen gorillas in the collection, although one adult male is on loan to Howlett's Animal Park. A baby had to be taken from the mother and is now in the zoo nursery. This is the first time a gorilla had to be raised from the first day in the nursery.

Ringling Brothers and Barnum & Bailey Circus had a very famous animal, Gargantua, which they tried unsuccessfully to mate. Gargantua finally died, and a second gorilla, Gargantua II, was obtained, which died a year or two ago. The body was brought up to the Yerkes Center for autopsy. When laid on the autopsy table, the animal was so large from back to front that the autopsy staff had to stand on boxes to reach the top of the animal's body to commence dissection. The dissection showed many changes in the animal, particularly in the heart and blood vessels, which indicated that it had reached a considerable degree of senility.

In 1961 the Phoenix Zoo bought two gorillas which had come to the United States from the Cameroons in Africa. They were a pair, a female, Hazel, and a male, Mongo. They settled down and were very happy in their home in the Phoenix Zoo. Then in 1969 Mongo died. Hazel, a very attractive and coquettish gorilla, weighing only 240 pounds, was left a widow. The zoo after a time purchased another male gorilla for Hazel. He was called Baltimore Jack, and they paid $5,000 for him. He is described by the society page of the Phoenix *Gazette* as the $5,000 wonder boy of the ape world, 6 feet 3 inches tall, eighteen years old, and about 300 pounds.

There was interest, excitement, and amusement in Phoenix when it was learned that Baltimore Jack would be flown from Baltimore Zoo to Phoenix in Hugh Hefner's DC-9 jet, presumably complete with bunnies. At that time a traveling evangelist was holding a revival in Phoenix. His name was Dr. Jack Hyles, and it is difficult to understand the mental processes of such a person who questioned the Christian morals of bringing a male gorilla in such a vehicle. Taking this up in what I hope was a sense of fun, the Phoenix *Gazette* said, "Will Baltimore Jack's morals suffer after riding in Hefner's plane? Not at all, say Phoenix Zoo officials, Jack will be sedated and strapped down the entire trip. He won't even see any bunnies or bunny-style stewardesses or whatever you call them that Hefner has aboard."

The entertaining affair was described in a most amusing fashion by Nicholas von Hoffman in the Washington *Post* of April 1, 1974. He said that the wedding between Baltimore Jack and Hazel was the biggest thing that happened in Phoenix since Goldwater got nominated for the Presidency and that a jeweler in the town produced two eighteen-karat gold rings in the form of miniature bananas for the wedding. A special wig was donated by Hair of Scottsdale. The best photographers in town elected to take the wedding pictures, and a Phoenix dentist said he would provide a wedding cake. The Phoenix *Gazette* said that the dentist who made this offer "was inspired in the suggestion by a wedding ceremony for two rhinoceroses. While at a dental convention, he attended the rhino wedding at a lion farm in West Palm Beach, Florida, last month."

The marriage between Jack and Hazel took place in July, 1970. By December the Phoenix *Gazette* had this to say: "Their

113

much desired consummation has not yet occurred. . . . Phoenix Zoo Director, Jack I. Tinker, predicts romance will bud for them next spring when it's warmer and cozier."

A headline a month later said, NO STORK ON WAY FOR GORILLA, TESTS SHOW, and the zoo director said, "What people have to understand is that Jack has never before seen another animal, let alone another gorilla. He's got a lot to learn."

Nicholas von Hoffman made an amusing comment: "The truth was Jack, like many another male who hangs around pickup bars and Playboy Clubs, could talk better than he could perform."

Another paper in the area, the *Arizona Republic,* had an article about the possible use of hormones and aphrodisiacs and went on to say that Baltimore Jack had not received any sex education and made this comment: "Any human locked in a cage during the formative years and through puberty, without parents and without examples, would probably have difficulty in becoming a parent."

Then Jack fell ill and died, and once more Hazel was a widow. On October 19, 1972, Baltimore Jack was preserved in formaldehyde and put in the arts building at the Arizona State University at Tempe.

The Phoenix *Gazette,* after a suitable interval, requested someone to provide a virile young stud for Hazel. "The possibility of the Phoenix Zoo showing pornographic movies to Hazel to stimulate romantic inclination has prompted much conversation around town—the London Zoo developed a film (entitled 'Wham, Bang, Thank you, Ma'am' and running 30 seconds) with the idea that gorilla see, gorilla do."

Hazel still managed to hit the headlines, and the *Arizona Republic* announced, GORILLA WILL TAKE THIRD HUBBY. According to van Hoffman, Hazel was to be sent to the San Diego Zoo to have a go at group sex, or communal marriage, with two other females called Alvila and Dolly and "a tall, $10,000, 400-pound" gorilla known as Troup, which had been donated to the San Diego Zoo by a San Diego newspaper. Apparently Hazel is still in San Diego, and according to van Hoffman, the last information the Phoenix viewers had about its gorilla was "monkey business just fine with Hazel and her friend." The latest news is that Hazel has a baby.

The Oklahoma Zoological Society in its publication, *Zoo Sounds,* had a fascinating story in its October, 1973, issue about the attempt to get its male gorilla to breed. The story, entitled

"Return to M'kubwa, or the True Love Story of Three Mountain Gorillas," is as follows:

Long time bachelor, M'kubwa, has finally met his matches (plural)! September 28 marked the second month anniversary for M'kubwa, his wife Josephina and his companion, Sumaili. Breeding activity has been reported for Josephina and is expected soon for Sumaili, who was still too shy with M'kubwa when she first cycled.

The bells started tolling for M'kubwa in May, 1971 when Josephina arrived on loan from the Tel Aviv Zoo, Israel. This first romance was short as M'kubwa proved himself to be a bully and alternately neglected and abused his wife to the point that she developed ulcers on her feet. The zoo staff ordered a separation for Josephina but allowed visiting for the troubled newlyweds through the bars between their cages.

The separation and subsequent recuperation of Josephina was interrupted slightly by the arrival of a new woman in M'kubwa's life. Sumaili, a twenty five year old female mountain gorilla from the Bronx Zoo, arrived on a breeding loan in July, 1972 and promptly built a fire in the heart of M'kubwa.

Rumor has it that ten year old Josephina took Sumaili aside and explained the whole situation of M'kubwa to her. Shortly thereafter Sumaili was introduced to M'kubwa when she was in estrous.

M'kubwa was obviously quite excited by the prospects of Sumaili, so excited in fact, that instead of a sweet kiss and nibble on the neck, he nibbled so violently that the two were separated so that Sumaili's wound could be treated. We never saw the kiss, if it occurred at all.

In our lowland gorillas, the male has been so dominant to a single female that no breeding has occurred. When two more females were added, the male learned some manners quite hastily as the three lowland gorilla females had conspired together to upgrade their social position at the zoo and downgrade the male, chauvinistic gorilla. Since then, breeding reports steadily increased, and one baby has been born, unfortunately premature.

With the evidence of change in the lowland gorilla male, zoo staff, working as marital counselors, recommended that M'kubwa be authorized two wives concurrently who could then work together to subdue the 500 pound wife-beater, M'kubwa.

For the last year Sumaili shared their half of the Gorilla Breeding penthouse on the lakeshore and, although no gentlemen callers came, they did establish a stable relationship together as wives-in-law.

115

While the two women gossiped away the time together, M'kubwa simply sat and waited with great patience, as he knew he would get his wives back, and it was only a matter of time.

In the meantime, however, the zoo staff was working diligently to devise a plan to make the threesome marriage work. Estrous cycles were recorded by the animal technicians and a date was projected on which both females would be neither receptive to M'kubwa nor would he be overly attracted to them.

A tranquilizer, disguised in a banana, was given to M'kubwa and the proper dosage determined which would dilute his aggressive nature to a low level. The zoo staff planned to keep him sedated for at least a month. In this way, he could approach the females in a proper manner instead of wild rampage when they were in estrous.

The final bells tolled for M'kubwa on the 28th of July when all doors were opened and all systems were go. The zoo staff anxiously watched and waited as M'kubwa sauntered into the females' suite.

The females apparently sensed M'kubwa was diluted and throwing caution to the wind promptly set about teaching M'kubwa a few manners. The fists and fur flew, punctuated with a few love bites and M'kubwa realized that either he had better shape up and act right or the ladies would promptly ship him out.

After a month . . . M'kubwa . . . settled down and accepted a less dominant role in the running of the gorilla building. His tranquilizers [had] been reduced to a very low level and it [appeared] to all concerned that a happy ménage à trois [had] come to pass.

The San Diego Zoo has had gorillas for quite a long time and has also been very successful in breeding them. Probably their most famous gorilla is Albert, a beautiful great male silver back. Albert is twenty-six years old now. When he was sixteen, in 1965, according to L. S. Nelson, who wrote about him, he learned to sing a love song which presumably was entitled "Gorilla, My Dreams." The reason for this was that a sweet young female gorilla from Africa had been moved into his quarters. Her name was Vila, and she had arrived in January, 1964. She was only six and a half at the time, and she had come from the ape island in the children's zoo and from the Primate Behavior Laboratory at San Diego. Albert had actually been a long time in the zoo. He arrived there from Africa in August, 1949. Two little female gorillas had arrived with him. This was the first time that any zoo had had the chance to raise three gorillas

116

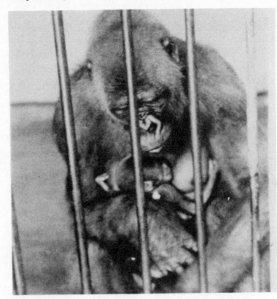

Kathryn nurses Frederika. (Courtesy of Jim Ellis and Oklahoma City Zoo.)

Gorilla baby at San Diego Zoo. (Photo by Ron Garrison, San Diego Zoo.)

Gorillas at San Diego Zoo. (Photo by G. Bourne.)

together. Of course, a little male with two females were a family group, and at that time there had been no gorillas born in captivity. Albert was the youngest of the three and could hardly sit up at that stage, weighing only ten pounds, and must have been little more than four or five months old. The females were at least six months older than he was. One of them was Bata; she was the bigger of the two. She was well enough advanced that she could hold her own nursing bottle of milk, but later, when the teeth began to erupt, it was estimated that she was probably younger than the other female, who was called Bouba. The three babies were not related to one another. They had been captured in different parts of West Africa. The three of them received a good deal of affection and cuddling, and they had some typical baby infections. On one occasion they had a nightmare which lasted two weeks, caused by some bug that they all got which caused them a high fever. They struggled for breath, their lips were swollen and blistered, and they were unable to eat solid food during this period and had to exist on little more than fluids.

When they were transferred to the sunny playroom in the zoo hospital, Albert was beginning to move uncertainly along on his arms and legs, following the little females. They would often come running to him and knock him on his back. He would get upset by this, reorient himself, and start off after the females once again. As he got older, he outgrew the females very quickly; of course, everything then was reversed. He showed all the characteristics of a small boy in liking sweets and disliking anything that was nourishing for him. He did not like strained peas and carrots at all, and they had to be mixed with apple sauce before he would eat them. He did not like milk. The three of them were soon taught to use a spoon and a cup and also a bowl. This was done primarily to see if they had the manual skills necessary to enable them to use these articles. Of course, they had no problem in learning to do this. One of the attendants working with these animals, Edalee Orcutt Harwell, was the first to persuade Albert to use a spoon by offering it to him filled with a sweet, orange-flavored vitamin preparation. There was no problem getting him to accept this and also to accept the use of the spoon and learn how to use it properly.

Albert soon became the number one animal in the group, and when Bouba began to scream and throw a tantrum, he would go over and cough a warning at her that would stop her screaming and send her running for safety. This is a very interesting fact because when females misbehave in the wild, this is pretty much what the number one male does. He goes over and coughs and makes some noise at them, and they immediately stop their misbehavior.

After another three years the three animals became adults, but that did not produce any offspring. It was the gorilla pair in Columbus, Ohio, that produced the first baby gorilla born and raised in captivity, in 1956.

Bata eventually went to the Fort Worth Zoo in Texas, and she was replaced by Vila. Albert tried to impress Vila by strutting and rushing around. The two of them remained together for quite a long time and copulated a number of times. Then on November 23, 1964, another young male, Trib, was added to the group and appeared to become well integrated. On June 3, 1965, a four-pound eleven-ounce baby was born to Albert and Vila. She had been the seventh gorilla conceived and born in captivity anywhere in the world, and she was called Alvila.

It might be of interest at this point to record the gorillas that have been born in captivity up to 1965. The first was a female called Kolo, born on December 22, 1956, at the Columbus Zoo, her parents being the Baron and Christina. The second gorilla to be born in captivity was Goma, a female, at the Basel Zoo, on September 23, 1958. Kolo was 1.87 kilos; Goma was 1.82 kilos. Stephi and Achilla, who produced Goma, also produced Jambo, on April 17, 1961. Tomoka was born to Nikumba and Moka in Washington, D.C., on September 6, 1961. There was a long hiatus of three years before the same pair produced Leonard on January 10, 1964. On June 1, 1964, Stephi and Achilla, in Basel, produced a third baby, Migger, who was a male and remained with the mother. Migger was 3.95 kilos at eighty days. On June 3, 1965, the San Diego Zoo had its baby gorilla, Alvila, which weighed 2.14 kilos, the parents, of course, being Albert and Vila. The Frankfurt Zoo in West Germany produced Max, a male, born June 2, 1965. He weighed 2.1 kilos and was the son of Abraham and Makulla. So it looks as though 1964 and 1965 were bumper years for gorilla production in captivity. Apparently the breakthrough for breeding gorillas in captivity had at last been achieved. There have now been, according to Ronald Nadler of the Yerkes Center and Marvin Jones, an expert on ape classification, 121 gorilla births in captivity, and 96 of these were viable infants.

Dr. Duane Rumbaugh wrote an article in 1966 on the first year of the life of San Diego's first gorilla baby, Alvila. Dr. Rum-

119

baugh at that time was professor of psychology at San Diego State College and a member of the Zoological Society's Research Council. Subsequently he became associate director and was in charge of behavioral research at the Yerkes Primate Center; he is now a professor and is chairman of the Psychology Department at Georgia State University in Atlanta. He noted that during the first week Alvila, at the age of four days, was able to lift her head from the mattress even though her body was prone. She also gave a startle response to a strong stimulus provided by slamming a door. Alvila was able to crawl along her crib as a result of powerful thrusts alternately of the right and left legs. If she was disturbed when she was sleeping, she would yawn and frown and wrinkle her brow in a very marked fashion. When she was beginning to feed, she would approach the nipple with her mouth stretched wide open. She would then run her gums and lips along the length of the nipple and only then begin to feed. Though she was strong enough to support the forepart of her body for a second or so on her fists with her arms extended, her muscular development was stronger in her hips and legs than in her arms. She had a rudimentary ability to orient herself toward a stimulus, for example, a bright light; but unlike the gorilla at Jersey, she did not make any effort to track a stimulus as it was moved in an arc about her. When she was in the supine position, she would be agitated unless she was given a blanket to cling to. After the first week she became less agitated under these circumstances. She could roll over from her stomach to her back by pushing with an arm, and she did not use her feet to do this in any way that could be observed. The first few days her body hair grew very rapidly.

During her second week her general body strength increased, and she became sensitive to quite a number of different sorts of stimuli. If she was crying and a door was opened, she would stop. If Dr. Rumbaugh snapped his fingers, she would stop all general activity, but she still had difficulty in fixating visually. There was no evidence at that time that she recognized any particular person, but she did begin to orient herself to the face of an attendant who cuddled her. By the end of the second week she would sometimes track a pen light in a room which had been darkened, but when the lights of the room were on, she did not track a red plastic square. In due course she was wearing diapers, and from her vocal complaints it was obvious that when the diapers were messy or wet, she was uncomfortable. On the ninth day she gave her first full gorilla scream with a wide-open mouth. On her eleventh day she sat alone without assistance for the first time. She did this by grabbing the bars of the crib and pulling herself up and holding onto them.

At two months Alvila was able to fixate visually and track a number of things which were brought into her visual field. She had been eating solid foods, but frequently refused new foods when they were first introduced. Although she was, even from the first week, able to roll from her stomach onto her back, she was not able to roll from her back onto her stomach until the end of the second month.

When she crept, Alvila used her arms alternately in addition to her legs. At seventy-one days Alvila's upper-middle incisors cut, and that gave her a total of four teeth. At two months and three weeks Alvila's lower-right lateral incisor cut, giving her five teeth, and the three lateral incisors showed signs of erupting in the near future. At three months Alvila had eight teeth: medial and lateral incisors, both upper and lower. She crept about in her crib a great deal. Later after a few days her creeping developed into a crawl and then into a very vigorous quadrupedal walking and running.

It was twelve months before Alvila was able to walk at all in a true bipedal fashion, which was very different from a baby gibbon with which she was compared and which was able to walk bipedally much earlier.

Sometimes Alvila would lie on her back and flail her limbs in a vigorous fashion as though exercising them. Only very strong stimuli, produced by slamming doors or banging hammers, would really startle her at this age. She was also likely to move very quickly from one emotional state to another. At one moment, for instance, she would scream, and the next moment she would be playing vigorously and perhaps the next moment would fall asleep. She also appeared to like association and contact with people.

In 1968 the San Diego Zoo obtained a small lowland gorilla called Masa. He had lived deep in the sultry jungle of the northeastern part of the Gabon and was captured in the spring of 1967. A Peace Corps volunteer, Craig Kinzelman, who was in the jungle with a group from the Bengum tribe, found the little baby close to the border of the Congo a little while after his mother had been killed by African hunters. Kinzelman described vividly his journey back to camp after nightfall, holding Masa in his arms, and how the little gorilla was terrified by the flare of the torches that the Africans used to provide some kind of light to enable the party to hack its way through the almost impenetrable jungle on its way back to the camp. Incidentally, "Masa" is the word that the Bengum use for gorillas.

Kinzelman took Masa to Dr. Schweitzer's hospital in Lambaréné and asked Mrs. Rhena Miller-Schweitzer, the daughter of the late Dr. Albert Schweitzer, to look after him. Masa became popular immediately with the hospital staff and patients and was allowed a good deal of freedom and spent a lot of time climbing about in the trees around the hospital and on the roofs of the buildings. He continued to have a very good-natured and affectionate personality, according to Dr. George H. Pournelle of the San Diego Zoo, who described Masa's early life. He had this good-natured, affectionate personality and retained that, but it was evident that after a while his carefree days at Lambaréné were coming to an end, since he was getting too big and powerful to have around the hospital. It was suggested by Craig Kinzelman that he go to the San Diego Zoo. Mrs. Miller-Schweitzer agreed with this, and on July 28, 1968, Masa, now weighing forty-two pounds and accompanied by Mrs. Miller-Schweitzer, came to the New World at the San Diego Zoo. The estimate of his age was about two and a half years. It had been thought that they would place Masa on special exhibit with Alvila, who had been born at the San Diego Zoo on June 3, 1965, and had spent most of her life in the children's zoo. She had outgrown the children's zoo and was transferred to something more substantial. The two of them were introduced to each other, and although Alvila weighed sixty pounds against Masa's forty-two pounds, they got on together very well, but Masa soon showed himself to be the head of the family. Alvila seemed to accept this without any particular problem. The babies were too young to produce offspring at that time, but the hope is that in a few years they will be able to produce a second-generation, captive-born gorilla for the San Diego Zoo.

An interesting historical note was found in an issue of the *San Diego Zoo News*, published at the end of 1950, and has the following comment concerning the year's activities: "January 19, Dr. and Mrs. Robert M. Yerkes arrived today to begin a preliminary study of the three lowland gorilla youngsters, Albert, Bata and Bouba. The Yerkes' chief objective during the current visit is to outline a study program which is to be followed throughout the gorillas' residence in the zoo. At the end of the first day's contact, it is difficult to be sure whether the Yerkes are more impressed by the gorillas or the gorillas by the Yerkes."

In January, 1974, the Houston Zoological Gardens completed and opened a beautiful new building for their gorillas. It is probably the nicest indoor facility for gorillas in the United States. The exhibit area in which the animals spend about ten to twenty-two hours a day is approximately 2,000 square feet in size. It is composed of several terraces at various levels and includes a shallow, cavelike area. There is a stream with four pools running through the complex. There are two swinging vines, three concrete trees with branches which go up to the ceiling, and a six-foot stump. The platform trees and rockwork are colored to simulate the actual appearance of a jungle area and receive sunlight through several rows of Plexiglas skylights. In addition there is a profusely planted area with several species of African birds and a fifteen-foot waterfall protected by a

Gorilla habitat display (1974), Houston Zoological Gardens. Hanging vines are utilized often in play behavior. (Courtesy of the director, Houston Zoological Gardens.)

Thirty-five-year-old gorilla in Central Park Zoo. (Courtesy of Dr. Ronald Nadler.)

Gorilla Gori at the Center for the Acclimatization of Animals in Monaco beats his chest in characteristic gorilla fashion. (Courtesy of Prince Rainier of Monaco.)

dry moat. These add color, life, and activity to the display. The walls of the exhibit are artistically painted with scenes of tropical vegetation. A public area is separated from the gorillas by an open dry moat similar to the one of the Cheyenne Mountain Zoo in Colorado Springs. A closed-circuit-TV surveillance camera continually monitors the public to ensure its proper behavior. Another camera surveys the exhibit area and the gorillas. Richard Quick, curator of the mammals, describes with justifiable pride the new development of their zoo. He said that in such an environment the gorillas become very active and utilize every area of the facility; they particularly enjoy the shallow pools and swinging vines.

A great gorilla killed on Mount Mikeno, Kivu District. From Ben
Burbridge, *Gorilla* (New York: The Century Co., 1928).

6

Learning About Gorillas Through the Sights of a Gun

IT is only in the last century and a quarter that we have obtained any real knowledge of the gorilla, though most of the knowledge has been obtained through the sights of a rifle. Before that a few legends and myths emerged from "darkest Africa," but it was impossible to tell if they referred to gorillas or to chimpanzees or even to some strange human inhabitants or whether they were simply figments of someone's imagination.

Perhaps gorillas were referred to in the Bible. In II Chronicles 9:21 there is a statement: "For the king's ships went to Tarshish with the servants of Huram; once every three years the ships of Tarshish used to come bringing gold, silver, ivory, apes, and peacocks."

The term "ape" was probably not used in Chronicles in the sense that we use it today (for gibbons, chimpanzees, orangutans, and gorillas) and may have referred to any type of primate, but of course it is possible that the ships did bring chimpanzees or gorillas, probably as babies.

It is highly improbable, however, that anyone except the people who lived with them had ever seen an adult gorilla until the nineteenth century—probably Du Chaillu was the first; he was certainly the first white man to do so.

In the temple of Juno in Carthage about the year 450 B.C. were deposited some tablets which described a voyage in 470 B.C. along the west coast of Africa by a Carthaginian admiral called Hanno. The tablets were known as the *Periplus,* and they carried a passage, now often referred to in works on the gorilla, which is of special interest to "gorillaphiles":

126

On the third day after our departure thence, having sailed by those streams of fire, we arrived at a bay called the Southern Horn; at the bottom of which lay an island like the former, having a lake, and in this lake another island, full of savage people, the greater part of whom were women, whose bodies were hairy, and whom our interpreters called "Gorillae." Though we pursued the men, we could not seize any of them; but all fled from us escaping over the precipices, and defending themselves with stones. Three women were, however, taken; but they attacked their conductors with their teeth and hands, and could not be prevailed on to accompany us. Having killed them, we flayed them, and brought their skins with us to Carthage. We did not sail further on, our provisions failing us.

This is a translation from the Greek by W. Falconer and was published in 1797.

Some experts believe Hanno did not, in fact, see gorillas because he had not sailed far enough along the west coast of Africa to reach their habitat, having gone only as far as Sierra Leone. It is likely, however, that 2,000 years ago gorillas had a much bigger range in Africa than they have occupied in recent history.

Was this the earliest possible record of the sighting of the gorilla by civilized man? According to Ashley Montagu, the word "gorilla" is not one which can be found in any African language known today, but the word "engena" is used by some natives for "gorilla" and "engeco" for chimpanzee. He believes that with the nasalizing and glottalization of the word "engena," it might easily have been mistaken for "gorilla" by Hanno's interpreters.

Du Chaillu believed that the Carthaginian expedition reached the mouth of the Fernan-Vaz River. Having originally thought that the animals which Hanno saw were chimpanzees,

> I now think it more likely [he wrote] that the gorilla was the animal seen and not the chimpanzee, which is generally less gregarious, and is not often found near the sea coast. As to the theory that Hanno's hairy men were some species of baboon, I think that very unlikely; for why would the Carthaginians hang the skins in the temple of Juno on their return to Carthage and preserve them for so many generations, as related by Pliny, if they were simply skins of baboons, animals so common in Africa that they could scarcely have been considered as anything extraordinary by a nation of traders and travellers like the Carthaginians.

Ashley Montagu gives two reasons for believing that Hanno and his party saw no gorillas: First, gorillas do not run away when chased and throw stones at the people pursuing them. As he says, either they run off or the big male makes bluffing rushes to hold the enemy at bay while the rest of the group escapes. Second, any human who got close enough to handle an adolescent or mature female gorilla would likely be massacred since even a female gorilla of that size is immensely strong. If the animals had been gorillas, "all the scampering would have been done by Hanno's party."

We will never be able to decide now whether Hanno and his party really did see and capture gorillas, chimpanzees, or any other kind of primate—human or nonhuman.

Another suggestion that the gorilla was known to Western man before the birth of Christ comes from the interpretation of figures as gorillas on two ancient gilded silver bowls found in Cyprus which combine Egyptian and Assyrian motifs. These bowls seem to be of Phoenician-Cypriot craftsmanship and date to the period following the downfall of the Minoan-Mycenaean civilization, i.e., about 1100 B.C. One of these bowls is in the Metropolitan Museum of Art, and Dr. Montagu had an opportunity to examine it personally. He found it impossible to conclude with any degree of certainty that the figure reproduced on this bowl was a gorilla.

Not long after Hanno's voyage, Aristotle published his famous *History of Animals,* and in it he referred to "the ape, monkey and the baboon." "The monkey," he said "is a tailed ape." He gave a detailed description of what an ape looked like. He may, of course, have been referring to a Barbary ape, which at that time probably ranged over quite a considerable part of North Africa and is still found there. It is the "ape" which has become famous as a simian occupant of the Rock of Gibraltar. It is not in fact a true ape at all, but is a tailless monkey. It has certainly been very well known around the Mediterranean since before the time of Christ. It was dissected by Galen as a substitute for the human body. At the same time there is still the possibility that Aristotle may have seen some anthropoid apes.

A reference in Sir Richard Burton's translation of *Arabian Nights* to a "gorilla lover" suggests that the gorilla may have been known in the Middle East about A.D. 5 or 6.

Two thousand years after Hanno the gorilla may have been seen by Andrew Battell, who had been captured by the Por-

A wild mountain gorilla rushes out of the jungle, beating his chest, seen by a member of the Martin Johnson expedition. From Martin Johnson, *Congorilla* (New York: Brewer, Warren & Putnam, 1931).

Gorilla and chimpanzee happily share a bed in Nairobi. From Martin Johnson, *Congorilla.*

The Martin Johnson expedition camped near Mount Karisimbi in the vicinity of Carl Akeley's grave. This is typical gorilla country. From Martin Johnson, *Congorilla.*

The grave of Carl Akeley, the man who was responsible for the establishment of the Albert National Park for the protection of mountain gorillas. From Martin Johnson, *Congorilla*.

tuguese. He encountered two types of animals which he regarded as apes, one of which he described as a "Pongo," and there is a good possibility that this was a gorilla.

In 1774 Lord Monboddo, in a famous work entitled *Of the Origin and Progress of Language,* reprints a letter from a man who had been captain of a ship which had traded along the coast of Africa. The letter refers to a creature known as an "impungu" and compares it with a "chimpenza." It seems likely that the latter was a chimpanzee, but was the "impungu" a gorilla? In the letter it is described: "This wonderful and frightful production of nature walks upright like a man; is from seven to nine feet high, when at maturity, thick in proportion and amazingly strong. . . ."

Then in 1819 Thomas Bowdich published *Mission from Cape Coast Castle to Ashanti.* He talks about a number of apes in the area of Africa known as Gabon. He called them "ingenu" and described them as being five feet tall and four feet broad.

The three references, Battell, Monboddo, and Bowdich, are the only indications of the possible existence of the gorilla between the time of Hanno's expedition in 470 B.C. and the identification of such an animal from skeletal remains in 1847—more than 2,300 years.

Up to 1846, despite stories of the existence of the large and ferocious animals eventually known as gorillas, there was in fact no definite evidence that they existed at all. In that year, however, the Reverend Mr. Savage visited his colleague, the Reverend Mr. Wilson, at the Gabon River in West Africa. In Wilson's house he saw a large skull which belonged to an animal with which he was not familiar. It was said to have come from an animal like a monkey, but very big and very ferocious. The two men eventually collected a number of these skulls and sent them off to Sir Richard Owen, the distinguished English anatomist. They also sent some skulls to the American anatomist Jeffries Wyman. Savage's description of the gorillas, presumably pieced together from Africans' stories, was as follows:

> They are exceedingly ferocious, and always offensive in their habits, never running from man as does the chimpanzee. . . . It is said that when the male is first seen he gives a terrific yell that resounds far and wide through the forest—something like Kah-ah! prolonged and shrill. . . . The females and the young at the first cry quickly disappear; then he approaches the enemy in great fury, pouring out his cries in quick succession. The hunter awaits his approach with gun extended; if his aim is not sure, he permits the animal to grasp the barrel, and as he carries it to his mouth, he fires; should the gun fail to go off, the barrel is crushed between his teeth, and the encounter soon proves fatal to the hunter.

This account was published by Savage in the *Boston Journal of Natural History* in 1844, and it was Savage who first used the name "gorilla" for the animal referred to by Africans and others at that time as engena. Some authors, however, claim that the name was first used by Professor Wyman. Savage's colleague, Wilson, gave the following account:

> It is impossible to give a correct idea either of the hideousness of its looks, or the amazing muscular power which it possesses. Its intensely black fur not only reveals features greatly exaggerated, but the whole countenance is one expression of savage ferocity. Large eyeballs, a crest of long hair, which falls over the forehead when it is angry, a mouth of immense capacity, revealing a set of terrible teeth, and large protruding ears,

altogether making it one of the most frightful animals in the world.

Another author, Henry A. Ford, also made his contribution to the gorilla mythology: "When he hears, sees, or scents a man, he immediately utters his characteristic cry, prepares for an at-

A gorilla shot by Captain Attillio Gatti. Trick photography was used to reproduce what Gatti said the gorilla looked like as it rushed him. From Attilio Gatti, *The King of the Gorillas* (Garden City, N.Y.: Doubleday, Doran & Co., 1932).

A study in hands and feet. From W. K. Gregory and H. C. Raven, *In Quest of Gorillas* (New Bedford, Mass.: The Darwin Press, 1937).

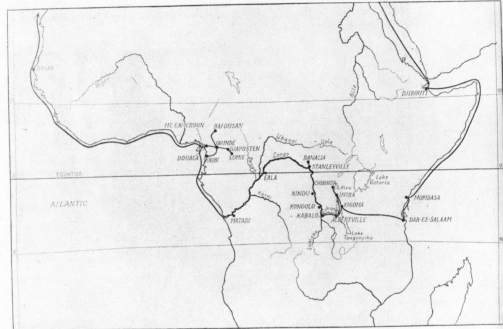

Route of the Raven expedition, from East to West Africa. From Gregory and Raven, *In Quest of Gorillas.*

tack; and always acts on the offensive. . . . Instantly, unless he is disabled by a well-directed shot, he makes an onset, and, striking his antagonist with the palm of his hands, or seizing him with a grasp from which there is no escape, he dashes him upon the ground and lacerates him with his tusks." Ford, who had also visited the Gabon, presented this interesting contribution to the Academy of Natural Sciences of Philadelphia. His story had many similarities to those of Savage and Wilson, and he also repeats the accounts of gorillas crushing musket barrels between their teeth. He did not, however, agree that gorillas would drive away elephants and did not agree that they built houses for themselves in the trees.

Mrs. Merfield and her baby, Trudie, beside a big bull gorilla. The great size of the animal can easily be seen. From Fred A. Merfield, *Gorillas Were My Neighbors* (New York: Longmans, Green & Co., 1956).

The scientist Carl Akeley said gorillas were misrepresented in museums. He referred to a bronze statue by E. Fremiet in the American Museum of Natural History. The gorilla itself is beautifully modeled, but over his right arm is what is meant to be a good-looking African woman, who in fact looks more like a Paris model. The gorilla is shown walking erect, and in his left hand he holds a large stone, presumably destined to be thrown at his pursuers. He also has been shot through the body with an arrow. The statue symbolized a number of inaccurate legends about the animal. The gorilla actually progresses on all fours with the hands clenched so that his knuckles bear the weight of the front half of his body; gorillas do not carry off women, and they do not throw stones at their pursuers. A number of other museums have mounted and stuffed gorillas in the rarely used erect posture; the famous evolutionist Professor Ernst Haeckel published a photo of the skeletons of man and gorilla, side by side, and both in the erect posture, and the equally famous biologist J. A. Thompson in his book *Outline of Science* shows a ferocious-looking gorilla walking erect.

A popular legend told about the strength and ferocity of the gorilla is that of the gorilla and leopard. As the story goes, the gorilla was walking in the forest and suddenly was met by a leopard. The gorilla stopped and so did the leopard, but the latter, being hungry, crouched and sprang at the gorilla. The gorilla roared and caught the leopard in midair. He grabbed him by the tail and whirled him round his head, but the tail broke off, and the leopard ran away, leaving his tail in the hands of the gorilla.

The famous biologist and defender of Darwin's theory of evolution, Professor Thomas Henry Huxley, had this to say in 1863 about the gorilla.

To the ordinary explorer or collector, the dense forests in equatorial Asia and Africa, which constitute a favorite habitation of the orang, the chimpanzee, and the gorilla, present difficulties of no ordinary magnitude; and the man who risks his life by even a short visit to the malarious shores of those regions may well be excused if he shrinks from facing the dangers of the interior; if he contents himself with stimulating the industry of the better seasoned natives, and collecting and collating the more or less mythical reports of traditions with which they are too ready to supply him. In such a manner most of the earlier accounts of the habits of the man-like apes originated; and even now a good deal of what passes current must be admitted to have no very safe foundation. The best information we possess is that, based almost wholly on direct European testimony, respecting the gibbons; the next best evidence relates to the orang; while our knowledge of the habits of the chimpanzee and the gorilla stands much in need of support and enlargement by additional testimony from instructed European eyewitnesses.

Huxley's remarks, made after Du Chaillu's visits to Africa, were very much to the point and were published about the same time as Du Chaillu's book; Winwood Reade's book, *Savage Africa,* in which he was very critical of Du Chaillu, was also published about then. Very little else happened, then in the 1890's Robert L. Garner went to West Africa. He was so intimidated by Du Chaillu's stories of the ferocious nature of the gorilla that he built himself a cage in the forest and sat in it day after day, waiting for the animals to pay him a visit. Very few did so and he saw very little of them, but with all his faults, Garner was the first person to get any idea of how the gorillas lived in the wild. Because he saw so little, he depended largely on the information he had got from the Africans about gorillas. In his writings he describes them as being polygamous. At first a male chooses a wife and remains married to her, showing some marital fidelity, then from time to time he adopts a new wife, but the old one is retained and so he eventually gathers around him a family consisting of seven wives and their offspring. These family groups were kept well separated from one another and often composed of ten to twelve animals. Any solitary gorillas, he said, were actually young males looking for a mate.

When 1900 dawned, the mountainous part of the Eastern Congo and Western Uganda became more glamorous than West Africa as a place to look for gorillas. In this area there is a great valley that stretches to the southern part of Lake Tanganyika from the Upper White Nile. This is the giant Rift Valley, which is about thirty miles long, and from this valley two masses of mountains rise to more than 16,000 feet on one side and to 10,000 on the other. For many years there had been rumors of large apes living in this area, and in 1866 David Livingstone during his famous walk passed through this area and talked of seeing Africans fighting with creatures which he referred to as gorillas. We do not know for certain that they were, but in 1890 Henry Stanley said that he believed that the gorilla lived in the northeastern part of the Congo. Then in 1902 a German officer, Captain Oscar von Beringe, traveled near the north part of Lake Tanganyika. He climbed the mountains in that area at an altitude of 9,300 feet, and he and his party saw a number of apes. This is what he said about them:

We spotted from our camp a group of black, large apes, which attempted to climb to the highest peak of the volcano. Of these apes, we managed to shoot two, which fell with much noise into a canyon opening to the northeast of us. After five hours of hard work, we managed to haul up one of these animals with ropes. It was a large, man-like ape, a male, about 14 meters high and weighing over 200 pounds. The chest without hair, the hands and feet of huge size. I could unfortunately not determine the genus of the ape. He was of a previously unknown size for a chimpanzee, and the presence of gorillas in the lake region has as yet not been determined.

The gorilla that von Beringe shot was in fact a mountain gorilla, and eventually it was named *Gorilla gorilla beringei,* in his honor. The mountain gorilla resembles the lowland gorilla very closely and it takes an expert to differentiate between the two, but there are actually thirty-four structural differences between

Gorilla country. (Courtesy of J. Sabater-Pi, Barcelona Zoo.)

the two animals. Because it lives in a colder climate, the mountain gorilla has longer hair and a thicker coat than the lowland gorilla. After von Beringe's discovery museum collectors and hunters swarmed the area, eager to collect as many specimens as they could. In a small 150-square-mile area in the region of the Virunga volcanoes fifty-four gorillas were shot between 1902 and 1925. Prince Wilhelm of Sweden led an expedition in 1921 which eliminated sixteen gorillas. Later on, between 1922 and 1925, the American hunter Ben Burbridge accounted for nine. In 1921 the American Museum of Natural History sent out a naturalist sculptor, Carl Akeley, to collect gorillas for them, and he killed five and also took 300 feet of film of the animals in the wild. This was the first film ever taken of wild gorillas. Akeley was very impressed with the animals and with the splendor of the setting in which he had found them. He pressed very hard for the establishment of a gorilla sanctuary in that area. The Belgian government eventually agreed to this, and in 1926 Akeley returned with his wife to survey the area for the establishment of the sanctuary. There, unfortunately, he died, and he was buried by his wife at the edge of the Kabara meadow high up on Mount Mikeno. Despite his death, the park was founded and was called the Albert National Park. It still exists as a sanctuary for mountain gorillas, and without it they might very well have been wiped out by now.

In the next ten years, from 1925 to 1935, there were two or three expeditions to Africa primarily to capture or to film or in some cases to shoot and kill a limited number of gorillas for various purposes—for museums mainly. One of the first of these was an expedition by Burbridge, who in 1928 published a book on his activities called *Gorilla: Tracking and Capturing the Ape-Man of Africa.* On one occasion when Burbridge arrived in New York with a young gorilla called Congo, he was surrounded by the press as he was describing how his left thumb and fingers had been damaged by the bite of a young gorilla when it was being captured. A reporter who was late in arriving came up and asked what the weight of an adult gorilla was. He was told that it was about 400 pounds. The story that eventually appeared in print about Burbridge's activities described him carrying out a hand-to-hand combat with a huge gorilla 400 pounds in weight. The article even reported that he choked the animal by pushing his fist down its throat. To call this story "unlikely" would be a euphemism; even if an adult gorilla were muzzled, a dozen Muhammad Alis could not have subdued him.

Ben Burbridge was a believer in the essential harmlessness of gorillas. On one occasion he was in East Africa camped on the slopes of Mount Karisimbi, listening to the Africans tell stories of the ferocity of the gorilla, and since he was hoping to persuade them to help him capture one alive, he protested that the gorilla was really a very harmless animal. The chief stepped forward at this stage and asked him if he really thought the gorilla was harmless, what would he say to the man whom he then presented? A figure limped forward, and on his shoulder, thigh, and knee were tremendous scars. The chief said that "this man went to the forest to cut a bamboo for a hut; a gorilla, unprovoked, attacked him." Naturally Burbridge heard only one side of the story. The gorilla's story might have showed that he acted in self-defense.

Burbridge made two expeditions to the Eastern Congo and claimed that his object was to study the gorilla, not to kill him, and also to learn if gorillas indulged in dances like those of the chimpanzee which Garner claimed to have observed and if the gorilla talked. Also, he had intended to take motion pictures of the gorilla and to try to capture a young one.

Lowland gorilla in the wild in Río Muni observes photographer. (Courtesy of J. Sabater-Pi, Barcelona Zoo.)

Young captive lowland gorilla receives a drink of milk in Río Muni.
(Courtesy of J. Sabater-Pi, Barcelona Zoo.)

Far up the Congo River, in the middle of the equatorial forest—practically in the center of Africa—are the ancient volcanoes Mikeno, Karisimbi, and Visoke, snow-covered in their upper reaches and more than 14,000 feet high. They are arranged in a triangle, in what Burbridge described as a guard of honor to the 400-square-mile area which was the kingdom of the mountain gorilla. To the north of these mountains is the Virunga range with its large peaks, and toward the southwest, at an altitude of 10,000 feet, is Lake Kivu. That this region is almost inaccessible and isolated may explain why mountain gorillas have been able to survive and raise their families for so many thousands and thousands of years.

For weeks on end Burbridge tracked mountain gorillas. Occasionally he saw them, but there were only fleeting glimpses, though these were enough to give him some idea of their habits. Gorillas, he observed, move around the forest in bands with a single adult male, females and young and some babies, and each band lives in its own particular part of the forest. The other bands rarely invade the territory of another band. Burbridge said, "That the adult male gorillas fight to a finish for mastery of the harem there is small doubt." He did not, however, see this; it was pure guesswork on his part, and there is in fact no evidence that it is so. Gorillas usually stay only one night in the camp that they make, and they build a number of nests close to the ground. On some occasions he found that the mothers with their young built large nests in the tall trees; "high" nests like these provided protection from leopards prowling in that area, and while a leopard would not attack an adult gorilla, he would quite likely snatch one of the babies if an opportunity arose. One African with Burbridge's party heard, one night, a fierce struggle in the forest which included the roars of the gorilla, and the next morning he found a dead leopard which had been badly mauled and apparently tossed to one side by the gorilla.

Burbridge was told stories of people being stolen from the villages by gorillas. Another anecdote was about an African who was killed by a gorilla when he was hunting; his heart was torn out by the animal, who carried it away to make medicine. There was also a story about a gorilla chief who was half man and lived in the jungle some distance away from where the Burbridge party was hunting, but all the other gorillas had to pay tribute to him.

On one occasion a group of gorillas came to a stop beneath the branches of a large tree on the edge of a forest clearing. Burbridge was hoping to photograph them, but his bearers refused to go any farther. Burbridge continued to walk toward the gorilla group, and the old silver-backed male and his family stood their ground, making hoarse growling sounds. He had only a .45 pistol, and when he got within twenty feet of the gorillas, they made a tremendous noise—not only vocalizing a lot but also crashing around among the vegetation. The old male gorilla roared, the young shrieked, and the females yelled and screamed. Now and again the old male would stop and beat his chest with his fists. Burbridge continued to press forward, closer and closer to the animals, then suddenly he found that they had gone; without the hunter realizing it, the animals had stolen noiselessly away.

Burbridge devised many tricks to enable him to get close enough to the gorillas to film them, but he had a lot of difficulty. Sometimes he would send men to the windward of them and hope that the animals would then drift down away from the scent toward his cameras. They could sometimes get gorillas out of the forest by imitating the snarl of a leopard or the cry of a young gorilla in distress. Sometimes they would beat on their own chests as if they were another gorilla band. One of the most successful tricks they used was to lead a party of, say, eight men into a thicket in full view of a group of gorillas, then only six of the men would come out and go away. Two would remain there with a camera. After the six had gone, the gorillas would assume that everybody had gone and then would come to the thicket to investigate. On these occasions the party got some very good film footage of these animals.

The gorillas never seemed to attack when they found two humans alone with a camera in the thicket. Since they are very curious animals, they were intrigued to find a white man fully dressed in clothes and looking quite different from the Africans, with whom they are familiar. This natural curiosity plus the intimidation of having the whirring motion-picture camera directed toward them were probably enough to prevent any attack.

Once Burbridge saw a gorilla, with his mouth open and his cheeks tightly drawn, beat on each cheek with the palms of his hands, producing a metallic and far-carrying sound. Another gorilla drummed on his chin, hitting the chin with the backs of his fingers. This resulted in a teeth-rattling sound similar to that which Burbridge had heard earlier, but had not been able to identify. On one occasion when Burbridge was filming a group of gorillas, they noticed him, hesitated, charged toward him, then saw what probably looked to them to be a one-eyed monster, Burbridge's camera, paused momentarily, terror-stricken in front of it, and then ran for cover. Once under cover they dispersed among the trees, observing the cameraman and

Capito de Nieve (Snowflake) plays with a human child. (Courtesy of J. Sabater-Pi, Barcelona Zoo.)

Snowflake and Muni play together. (Courtesy of Dr. Arthur Riopelle.)

his African assistant. In their excitement some gorillas began climbing trees, others began shaking young trees and dancing about, and all this tumult Burbridge was making into a unique film. At this point his gun bearer, Joe, plucked his sleeve and pointed; there was a large gorilla coming toward them. Occasionally he would stop to beat his chest or beat on his chin, producing a tooth rattle, or he would slap his cheek. He was thirty paces away when he lunged over to the trunk of a fallen tree and uttered a menacing roar that caused Burbridge to drop his camera and cover him with a rifle. He roared several more times and then beat his chest and then charged. As he charged, Burbridge fired above his head, and amazed by the noise, the gorilla stopped, turned, and fled into the forest.

Burbridge had some difficulty working out a plan to capture a young gorilla. He did not wish to shoot a whole group in order to get at the youngster or even to shoot a mother with a baby. He hoped to find some way to capture a young one without a massacre. While on this errand and tracking a group of gorillas, Burbridge momentarily left his trackers to take a short cut. The gorillas apparently backtracked and ran into the last of his group, and he suddenly heard a noise among his carriers; he came rushing back to find a mass of heaving human bodies. In the midst of them was a young gorilla that the natives were trying to hold down; Burbridge quickly slipped a sack over the gorilla's head, tied it securely, and the animal was captured, but not without cost. Some of his men had some small lacerations, and one had his knuckles crushed in a bite from the youngster.

The animal that it took all these men to subdue was only a twenty-two-pound gorilla. He was placed in a box and he spent the whole night beating on it in a rage. He refused to eat at first and was extremely savage, and then eventually, as he got hungry, he began to eat the forest foods that were brought him. Finally he was persuaded to sit on Burbridge's knee and drink from a cup. None of the black men, however, could establish any relationship with him. For some reason he remained consistently hostile to them. At the time that he was captured, this animal was the only gorilla in captivity in the world.

On another occasion a group of females with their young were pursued down the side of a canyon. Still giving chase, Burbridge managed to grab a young gorilla. But in doing so, Burbridge fell, splashing down into vegetation and water at the bottom of a canyon, and there he lay on the young gorilla and tried to subdue him. The gorilla, however, was more than a handful, scratching and biting, and since he weighed about sixty pounds, he was a real problem for Burbridge to handle. One of the Africans threw himself on the struggling pair, and eventually he and Burbridge managed to get the animal into the sack and secure it. Burbridge was disappointed that his other helper, Joe, had stood by and not offered his help in the struggle. As he turned to say something, Joe motioned to him to follow and led him to a trail of blood that disappeared into the undergrowth. During the fight a gorilla had charged the group and was almost on top of them when Joe fired a shot which averted what would have been a human disaster.

A much bigger gorilla, a young male weighing 126 pounds and called by the Africans Bula Matadi (which means "great

Wild gorilla studied by Dian Fossey. (Courtesy of *National Geographic*.)

master"), was also captured. It is said to be the largest specimen ever recorded even to this day. At one point when they were following a gorilla trail, a group of gorillas was seen feeding on a bed of wild celery. Following an audible command from the large male gorilla, the females with babies and the young males joined him, and when they were all assembled, the group started marching off. The humans kept very quiet, planning to pick up the trail later. While they were waiting for the gorillas to get out of sight, a female gorilla with two youngsters who apparently had not heard the male gorilla's order to march came walking through the celery beds and went into a thicket. They waited, hoping she would come toward them, but she didn't. So two of the Africans were sent upwind, in the hope that when the female got their scent, she would move away from it toward Burbridge. Eventually the mother and the two baby gorillas made a sudden rush through the thicket; Burbridge was crouching behind some boulders when the female landed on top of them. She then jumped to the ground and began to rush away, but passed Burbridge so closely that he found it impossible not to fling himself at her to try to bring her down. But the mother and the babies jumped clear of the rocks and ran into the underbrush, and then from behind came a young male gorilla that was passing him so closely that Burbridge tried to jump it also. He landed on the animal, clutched its throat, and hung on with desperation. Twice he tore himself out of the gorilla's teeth and left part of his clothing behind in its mouth. Several times he broke from the clutch of the animal as it was dragging his head and his throat toward its open jaws. Finally the gun bearer flung himself on the gorilla and the other Africans joined in, but the gorilla was still fighting "with the fury of a madman as he heaved and bucked under the weight of his enemies." Finally he was spread-eagled and his hands and feet were tied and he was eventually transferred into a sack. For fifteen feet around, the grasses and the young trees had been flattened by the struggle. Burbridge had lost one side of his hunting shirt. Both his hands had been mangled. He had one thumb that had been broken and crushed. He remembered that he had bitten the gorilla's fingers when they were in his mouth. He looked at the men and realized that the Africans' unclad bodies had suffered even worse than his. They were eventually able to chain the captured gorilla to a tree, but for three days he refused food or drink and remained very aggressive. Bula Matadi never reached civilization. A month after he had been captured he met a horrible fate. He was stung to death by ants.

Burbridge had been given permission by the Belgian government to capture eight young gorillas. Of the first four, only one survived and it was sent to the Antwerp Zoological Gardens. Later on four more were captured and two survived. One went to Antwerp, and the other, who was called Miss Congo, went to America. Burbridge, who later became concerned about the decimation of the gorilla population, estimated that there were about 2,000 gorillas in Kivu District and he recommended to the Belgian government that the present gorilla preserve be extended so that mountain fastnesses were included. This would, in effect, increase the gorilla sanctuary from 250 to about 500 square miles.

When Miss Congo arrived in America, she faced a battery of moving-picture cameras and newspaper reporters. At that time she was the only female gorilla in captivity, and with the death of other gorillas in zoos elsewhere, she eventually became the only one in captivity.

Eventually Miss Congo was bedded down in the country house of Burbridge's brother in Jacksonville, Florida, on the banks of the St. Johns River. Under the care of his brother's wife, the gorilla became strong and vigorous, developing in eighteen months from a 40-pound animal to one of 120 pounds. She was given a diet of Klim evaporated milk, apples, peaches, oranges, baked bananas, and baked sweet potatoes. On this diet she seemed to do very well. Eventually Dr. Robert Yerkes, founder of the Yerkes Primate Center, made the acquaintance of Miss Congo and carried out a series of detailed tests on her which formed the basis of a book he subsequently published, *The Mind of the Gorilla*. Miss Congo enjoyed the testing very much, and according to Burbridge, she was often so impatient to start her tests that "she would lead Dr. Yerkes to the door of her cage and push him out in order that he might arrange, without the enclosure, the necessary apparatus used." She was found to be able to use a stick to rake in food. She was shown an apple which was put into a pipe and pushed into the center of it—but it didn't take her long to realize that if she lifted up one end of the pipe, the apple would slide out the other side. On another occasion Dr. Yerkes hung an apple out of her reach and placed a stick near it, with the intention that the animal would use the stick as a tool to knock down the apple. She did not seem to be able to do this, but she did pile boxes on one another to get a reward.

She found an original way of using the stick. Instead of knocking the apple down with it, she placed the stick nearly vertically under the apple and then climbed on it and was able to grab the apple with one hand before the stick fell over.

Dr. Yerkes sums up his experiments with Miss Congo as follows:

By means of sticks, ropes, chains, bottles, boxes, a mirror, and other simple appliances, more than a score of novel problems were set for Congo. Most of them she solved eventually, some by what appeared like random action and the selection of profitable acts, others by observation of essential features or relations in the situation and immediate adaption. Evidences of psychophysiological processes in the gorilla are abundant and varied. Clearly trial and error as a description of adaptive procedure is incomplete and frequently inapplicable. Often there appear evidences of critical points in adaptive endeavor at which the nature of activity suddenly changes. Many of the objective characteristics in these suddenly achieved adaptations are observed in human ideational behavior. It, therefore, seems probable that the animal experiences insight. Various experiments proved that out of sight is not necessarily out of mind.

In addition to Akeley and Burbridge, a number of other people in the 1930's and 1940's explored the gorilla country. They included Jasper von Oertzen, Eduard Reichenow, Charles Pitman, Lucien Vlancou, and Fred Merfield. Fred Merfield spent fifteen years in the Cameroons collecting lowland gorilla specimens for museums and made a number of valuable observations on their life and habits. He was the first person to describe chimpanzees using sticks to collect food by sticking them down ant holes—an observation later confirmed by Jane Goodall.

A number of interesting gorilla legends were also told by his African guides to a hunter named Martin Johnson. One of these was that when a gorilla attacks a man, it will tear off his arms and legs and throw them away. An African guide told him the story of a gorilla which attacked a man with a large club and beat him to death. Another anecdote which circulates among the Africans is that when the leader of a gorilla pack becomes senescent, the gorillas in the group beat him to death; or alternatively, when a gorilla gets too old to hold his group any longer, he leaves the group and goes into the jungle and commits suicide. There is also a story in which the old leader is driven from the pack, and at night he comes back and kills each gorilla in the pack, one by one, finishing up by commiting suicide. This, of course, is a perfect way for eliminating an entire gorilla population, but it is only a legend. A report which reputedly came from a white man who had lived for many years in the haunts of the gorillas said that a pack of gorillas had slain two African women in the mountains, and the tribe from which the women came was now at war with the gorillas. This is the fabric from which gorilla myths are woven.

Johnson confirmed that Mikeno gorillas live almost entirely on wild celery and bamboo shoots and the tender buds of trees and bushes. Most observers of gorillas have noticed that they do not seem to drink in the wild, but they scarcely need to, for every morning the jungle is very wet either from the morning dew or preceding rain. The water that forms on the leaves and stems—particularly on the bamboo and wild celery—runs down and accumulates in the space where the leaves join together. This, together with the fact that these plants contain a considerable amount of water in their pulp, suggests that this is how the gorillas get all the water that they need. Gorillas in captivity, however, drink a lot of water.

Gorillas make nests, and Johnson and his wife ran across an area which had thirty distinct nests in it. They were situated in grass that was four feet high, and in building their nests, the gorillas had crushed areas of it down. Some of the nests were in contact; others were situated at varying distances apart. When building a nest, the gorilla will squat in a particular spot and then pull the surrounding grass down around him. He will then break up pieces of branches into smaller pieces and use this to lay on top of the grass that has been pulled in. In addition, Spanish moss and sometimes a number of bamboo leaves may be added.

Gorillas regularly soil their nests by defecating in them. The dung of a gorilla in the wild is rather like that of horses; it does not stick to the body, and defecating in the nest does not dirty the animal. The chimpanzee, on the other hand, produces much more fluid feces, and he never defecates in the nest, but always over the edge of the branch or the edge of the nest. The same thing applies to the orangutan.

When the jungle growth gets very dense, instead of trampling the grass and vegetation down to make their paths, gorillas tend to produce tunnels. Johnson and his group crawled on their hands and knees for a very long distance along some of these tunnels and were lucky they did not come face to face with a gorilla as they crawled along. In some cases the tunnels were as much as three feet off the ground. The apes apparently had no problem in moving along them, but the humans found themselves every now and again falling through the bottom. Johnson's primary objective was to get film of gorillas. The first one that he saw might be more appropriately described in his own words—particularly as they tend to reproduce the horror which surrounded and clouded the gorilla for so many years:

149

Suddenly, a huge black face appeared through the bamboo. It was then I found the source of the fabulous tales about this fearsome looking beast. The face was black, like oiled and polished leather; black as anything you will ever see. Framing it was black, close-cropped hair, through which round, small ears were peeping. Two eyes stared solemnly and directly at me. There was something about those eyes that suggested an evil spirit. They seemed to be glaring right through me as though some Satanic judge of the nether world were considering the penalty for one who dared invade his forbidden precinct. No wonder people believed these hairy creatures to be half man, half demon. That face with its curling, sneering lips, looked cold, cruel and murderous.

Having observed the camera and the cameraman, the gorilla broke into screams and was immediately joined by a number of other gorillas in the neighborhood. The noise they made sounded as though the demons of hell had broken through the earth's crust and were going to tear it apart and throw the pieces around the universe. Leaving his camera and running to a patch of grass where he had seen a black shadow, Johnson tells how he was suddenly confronted not more than fifteen feet away:

> An enormous gorilla slowly rose on his legs, grasping vines with his two black hands. He opened his enormous mouth and made the most blood-curdling yell ever heard, directly at me. I could see his red tongue and blood-red gums. Sword-like fangs were bared by the snarling lips; flanking them were teeth, huge and sharp. Had I not known better, I would have sworn that this ape was 10 feet tall, weighed 1,000 pounds, so vivid was the impression. And to this day, although I've seen many gorillas, I will swear that this snarling ape in the bamboo patch is the biggest of them all.

Johnson was more or less frozen on the spot and realized he had no gun or any kind of weapon he could use to protect himself. All the gorilla stories that the Africans had been telling came back, and he was anticipating being torn limb from limb, at any moment. In the middle of his terror the ape dropped down onto all fours, turned his back on him, and ran off into the woods. Though Johnson had been very close to being impaled on the tusks of a mad elephant and had stood in the way of rhinos thundering at him and also in the path of angry lions,

150

nothing was so etched into his mind as the confrontation he had had with that gorilla.

Later on they ran into yet another old silver-back male. (The term "silver back" is used because these older males tend to grow whitish hairs on their back which, mixed with the black hairs there, give them a silvery appearance.) One of their coexplorers, DeWitt Sage, was in the lead; behind him was Johnson and behind him his wife, Osa; following her was one of their African guides, Bukari. They were all moving forward when the old silver back suddenly screamed and made a rush at them and did not stop until he was within eight feet of the group. Sage, who was overcome by the sharpness and unexpectedness of the rush, fell over backward against Martin, who fell backward and hit Osa, who in turn, when she fell, knocked Bukari over. So they behaved exactly like a set of dominos, and the old gorilla must have really thought that his rush was worthwhile when he could produce this degree of chaos in the group of humans. The group was exploring in the Albert National Park sanctuary in the Congo, and it was understood that they would not use a gun unless it was necessary to protect their lives. So they did not go after the old gorilla, but retreated to decide the next step they would or could make. The fact that they retreated from the tunnel that they had been walking in encouraged the silver back to start a number of rushes down the tunnel, and he terminated these rushes just near the entrance, not very many feet away from where the group was sitting having a powwow. During the night the party was awakened by screams and yells and alarmed barks. In the morning when they surveyed the area, there were nests that the apes had obviously been sleeping in the night before, and throughout the area were footprints of two large leopards. They could tell from the footprints in the soft soil the direction in which the gorillas had gone. The leopards had apparently intended to sneak up on the gorilla group, grab a young one, and carry it off. When the gorillas were aroused, however, the leopards thought the better of picking a fight with them and ran for cover.

The gorilla expert George Schaller has described a fight between a gorilla and a leopard. While a group of gorillas was under observation, there were sudden terrifying roars from the males, and everyone rushed for cover. A leopard had come into the area, which was 9,000 feet above sea level, and leopards were rare at so high an altitude, but the cat had obviously followed the gorilla group. Eventually the male gorillas got into

151

Wild gorilla with Dian Fossey. (Courtesy of *National Geographic*.)

formation and, roaring at the tops of their voices, they charged the leopard. The leopard turned and ran. Later on one of the big silver backs became ill and left the group, accompanied by a black-back male the researchers called Junior. Junior seemed to be looking after the sick silver back. The two humans who were observing the animals, however, felt that since this animal was sick and had begun to show signs that he might die, there was a good chance that the leopard might come back. The next morning the humans returned to find that the old silver back and Junior had not returned to the group. Gil Clayton and W. Frazier tracked them and eventually found the silver back dead and Junior on all fours alongside him, presumably on guard. He suddenly showed excitement, reared up on his legs, and roared. Clayton turned to see that the leopard was slinking toward him. The leopard continued to advance despite Junior's intimidating roar, and Clayton expected Junior to turn tail and run, particularly since there were no other gorillas in the immediate vicinity to help him. But just as Clayton had lifted his rifle to shoot the leopard, Junior charged. Shortly after that the leopard leaped at the gorilla and they became locked together

Young captive gorilla with Dian Fossey. (Courtesy of *National Geographic*.)

in a fight to the death. The leopard was attempting to dig one claw into the gorilla's eyes while his rear claws were raking Junior's abdomen. They fell to the ground twice and rolled over. Neither would let go, and blood ran down the gorilla's face, chest, and arms. He struggled to his feet, still screaming, and grabbed the leopard's rear leg, wrenched him away, and threw him by the leg into the bushes. The leopard turned to attack again, and as he did Clayton fired two shots at him. Junior had fallen back on the ground and was bleeding freely from his many wounds. The leopard and Clayton gazed into each other's eyes for a moment, and then the leopard, merely wounded by the bullet, leaped onto Clayton's back, knocked him down, and sent his rifle flying into the bushes. The leopard was all over him. Clayton managed to get his left hand on its leg and pushed the animal back. He tried to get his knife out of its sheath with his right hand. He brought his knees up and put them against the leopard's chest in an attempt to hold him off. He finally managed to roll the leopard over, sit above it, and drive the knife into the thorax just below the rib cage. It took several stabs with the knife before the animal finally died. When Clayton got to his feet, bleeding badly, he found that Junior had crashed off through the bushes to return to his group. Clayton said it took about a month for his own wounds to heal, and by then he found that Junior was now in charge of the group and that his wounds had healed completely too. He was in charge because the old silver back had been the number one man, and now he was dead.

Johnson and his wife now set out to find Carl Akeley's grave in the saddle of Mount Mikeno, and Africans had told them that they had seen gorillas up there, some only a few miles away from the grave. The Johnsons found the journey very hard; they soon grew weary, and discomfort was added by the continual drizzle. To reach this height they then moved along for a few more hours and finally arrived at Carl Akeley's grave, which they found surrounded by a high stockade that had been erected to prevent the buffalo in the neighborhood from damaging it. When they examined the grave, they found that it needed some repairs. Some logs in the stockade had gone rotten and new ones would be needed. But the cement slab that covered the grave was intact. It took three days to repair the grave, and Johnson's wife, Osa, filled in all the crevices around the concrete slab which had been caused by the water and rain washing away the surrounding dirt. She had the Africans collect sod and place it around the grave to hold the dirt in place.

Sage and Johnson made a side trip while they were on the mountain, to the spot which Carl Akeley had described as "commanding the most beautiful view of Africa." Mr. Leigh, who had accompanied Akeley on a number of his expeditions, had actually made a number of his paintings here which later on provided the background for the gorilla groups which were established in the American Museum of Natural History. Johnson gives a lyrical account of this area, and perhaps it should be repeated here. They had been standing for some time waiting for the fog which regularly rolled in through the saddle of the mountain to disperse. Eventually it did, and they were able to see Lake Kivu in the distance. "The scene unrolled before our enraptured gaze was a magnificent panorama, noble, majestic and overpowering in its effect—a fitting canopy for the final resting place of Carl Akeley, who was its discoverer for the world of white men."

According to the local Africans, gorillas, although they have what appear to be well-developed chests, are very susceptible to pulmonary diseases. The Africans claim that they often can be heard coughing, especially at night, and they believe that many gorillas die from chest ailments every year. Johnson, by way of making a joke, suggested to his porters that the next time they came up there they should bring the gorilla some blankets, and much to his surprise, his audience of porters thought highly of this suggestion. One of them said, "Yes, Bwana, that would be a good idea. Then the gorillas could cover themselves at night and not catch cold." They thought the idea of mountain gorillas, in the future, wandering around wrapped in blankets was very funny. Perhaps some future explorer seeing this would be sure he had found a real missing link. The Africans were also greatly intrigued by the woolly gloves worn by the white members of Johnson's party because of the cold. The gloves created great hilarity and were called "gorilla hands" by the Africans. They had never seen anybody wearing gloves, and had never heard of them, and could see no reason for anyone ever to wear them.

On one occasion the party set the stage for filming gorillas and sent a group of porters around the back of a pack to frighten them so that they would dash past and through a clearing where the camera would be set up. Although several gorillas dashed past, none of them came through into the clearing where the camera was set up, but went past it into the jungle.

Another party set off with a number of dogs that would help them not only track down the gorillas but would also help to startle the animals. The African chief explained that the dogs had been specially trained to track down gorillas and would even hold a gorilla at bay while the hunters walked up and captured it. It was not long after they set off that the dogs began to howl, announcing that the gorillas were around even though they were a mile away. As they progressed, the dogs barked continuously. Thus the gorillas were given plenty of warning to keep ahead of the party. The chief insisted that this was the way Africans hunted gorillas; they would soon get tired, he said, and the dogs would then make short work of them. Finally the dogs, all excited, were unleashed to go in for the attack; but they immediately turned around and ran in the opposite direction from the gorillas, with their tails tucked between their legs and with the Africans in full chase after them trying to get them back on the leash. The dogs, however, were too shrewd to be released and had no stomach for this expedition at all. The dogs were eventually captured, but the confusion continued. One of the dogs barked, and the Africans who were in the lead of the expedition thought it was a gorilla close at hand. This time it was, and they turned around and rushed away in a state of abject terror. At the end of the day it was apparent that the party had been traveling in a big circle, and the net result of their filming was nil.

They caught two young gorillas by cutting down a tree on which they had taken refuge. One was a male and one was a female, and it was thought that permission might be given to take a pair of them out of the country. This was, in fact, the first pair of gorillas that had ever been captured in Africa. Each animal weighed more than 100 pounds, and at this weight they are immensely strong. It was quite a feat to capture them, especially without having to cause any harm to the animals. The gorillas were given water and corn and sweet potatoes, and they ate and drank very quickly after their capture, which surprised everybody and which was contrary to the experience of most other explorers up to that time. That night the camp was surrounded by gorillas who were calling to the young ones, who answered them back. This two-way conversation went on for hours. The captives beat their chests at intervals throughout the night, and there was a response of chest beating from the jungle around them.

There has been much speculation as to the nature of the

chest beating. Some scientists believe it is an outlet for surplus energy or else some kind of indication of nervousness. In the case just described it seems to be a recognition sign, and it is possible that when the animals are in the dense jungle, they do this from time to time to keep contact with each other. A silver back on Mount Mikeno has been seen to beat his chest as he looked curiously at humans, and at that time he was showing neither fright nor aggression. They have also been heard to beat their chests in the jungle when they were not even aware that human beings were present. Sometimes while a group is feeding, a gorilla will stand up and beat his chest. Sometimes when they are screaming and showing signs of fear, they will even stop their flight briefly to beat their chests.

At the Yerkes Center the gorillas seem to beat their chests with no particular relationship to anything, and they start it at a very young age. We have seen animals of two years of age beat their chests, and very often the young ones of about four to six, running around in their cage, appear to beat their chests seemingly from exuberance. The precise significance of chest beating is still unknown.

On one occasion an old silver back was found by Johnson's group during the day still lying in his nest. Upon the encounter, the animal jumped from the nest and ran crashing into the jungle. Around his nest were a large number of corn cobs. He had eaten most of the corn from some of them and even part of the cob itself. When the trail was followed to see where the corn had come from, it was found to lead to a garden that was enclosed by a log stockade, and the logs were set so close together that they touched each other. According to the African guides, this was done to stop elephants and buffalo from getting into the garden. The gorilla must have climbed this fence because inside ears of corn had been pulled away from the plants, but there was no sign of any of it having been eaten in the garden. Somehow the gorilla had managed to carry thirty to fifty fresh corncobs over the fence and then some distance to where he had piled them up around his nest preparatory to eating them. He could not have carried many, if any, corncobs in his arms while he crawled over the wooden wall, so he must have pulled them off the plants, thrown them over the wooden wall of the garden, climbed over, picked them up, and carried them to his nest. No one has ever seen a gorilla actually do such a thing, but this seems to be the most likely explanation.

Farther on in their travels the party was offered a small sick baby gorilla by an African, and they paid $60 for him. The animal was in a pretty poor state, but improved after a few days. The party finally arrived at the town of Irumu, where they called on the territorial administrator and asked him to send cablegrams to Brussels requesting permission to keep the two gorillas that they had captured. The one that they had purchased apparently did not need a government permit. They were authorized to keep the two gorillas. On their way down to the seacoast the Johnsons stopped at Nairobi and had the opportunity to meet Mrs. Akeley, the widow of Carl Akeley, who had accompanied him on all his expeditions. They bought an estate near Nairobi and spent four years there. Then they began to think it was time to return to America. They had a number of animals that they planned to take back: two chimpanzees, two young gorillas that they had caught, a colobus monkey, a white-nosed Congo monkey, a cheetah, and the young gorilla they had purchased. They journeyed by railroad to Mombasa and boarded the steamer *Njassa.* The steamer took them to Genoa, and they transferred for the journey to America to another ship, the *Excalibur,* an American Export Line steamer which took 120 passengers. They temporarily settled down in New York and quartered their animals in Central Park Zoo. They used to take their animals over to their penthouse from time to time for entertainment. On one occasion the young baby gorilla, Okara, after spending the night with them, let down the supports of their breakfast table, which resulted in the coffee and food on the table being spilled on the floor. He finished this activity off by rolling in the poached eggs on the floor. He was punished and sent up to the bedroom, where he had even more fun using powder and lipstick which he found. He then got into the bed, covered himself up with the blankets, and went to sleep. The two big gorillas stayed in Central Park Zoo until about 1931. Subsequently they were sent to the San Diego Zoo in California, where they had a large open-air cage with trees to play in.

Johnson said that now that he had his animals safely landed in America, he was sorry that he had brought them home. He was distressed to see his two beautiful gorillas imprisoned behind bars in a small space whereas only a few months before they had the whole Congo to roam in. He was also concerned for the cheetah which they had brought back with them when he realized that the animal was doomed to idleness and confinement for the rest of its life. Johnson, in the course of his writings, made some interesting comments about the intelligence of three major great apes and also of the lesser ape, the gibbon. He had already owned orangutans and chimpanzees. He was not of the opinion that any one of the apes—and presumably he was talking primarily about the three great apes—showed any great intelligence over the other. Their mental reactions, however, were very different, but the mental reactions of different cultural groups of humans are different too. Comparing them would be like trying to compare Edison with President Hoover, and Einstein with Lindbergh. Each of these men thought along different lines which required an individual standard of measurement. It was the same with the different species of the anthropoid apes. Johnson's chimpanzees were very quick at learning. For instance, if they saw him driving a nail with a hammer, they would soon get the idea and try to do it themselves. The following day they would have completely forgotten the incident, and for them to really learn anything would require constant repetition. The gorilla, on the other hand, particularly Okaro, was very slow to get hold of an idea, but once he had crossed the bridge, he never forgot what he had learned. The gibbon knew that there were some things it was allowed to do and some things it was not allowed to do, but if it did something wrong, it would hide away until it thought its owner had forgotten that it had disobeyed. On the other hand, if the orangutan did something it knew was wrong, it would not hide, but it would act in such a self-conscious and guilty fashion that even if you didn't know what it had done, you could be certain it had done something wrong.

When Captain Attilio Gatti was very young, a famous Italian African explorer, Captain Pisciscelli, who was a friend of his father, used to visit the house and regale the household with stories of his adventures. In one of his tales, as Gatti reminisced, he spun a yarn about a gorilla in the following manner: "But the gorilla, only slightly wounded, was more infuriated than ever. With the swiftness of lightning, he jumped upon the audacious Pygmy and with one squeeze of his powerful arms crushed him, just as you can crush an empty matchbox between your fingers."

When Gatti heard this story, he was only seven years old, but it made an impression which remained with him for the whole of his life and changed his ambition from that of becoming a trombone player in an orchestra to becoming a famous explorer, particularly an explorer who hunted the gorilla.

He did indeed turn out to be a first-rate explorer. At the end of the First World War, during which he served in a regiment of the royal house of Italy, he carried out many long expeditions in Africa. He crossed it from east to west, in fact, from Mozambique to Angola, and from north to south, that is, from Cairo to Cape Town. He was anxious to hunt the gorilla in Kivu in the Belgian Congo. Kivu is really the heart of Africa because it is situated just south of the equator, about 3,000 miles north of Cape Town. If you draw a straight line from the Indian Ocean to the Atlantic Ocean, you will find Kivu just in the center of that line, situated high in the mountains.

Prior to 1919 it was highly dangerous and extremely difficult to approach these mountains, but in 1919, when the Belgians occupied Kivu, the country around became more accessible. The gorilla of that area had been surrounded for many centuries by the Pygmies, who were very fond of gorilla meat and made war on gorillas without ceasing. With the arrival of Belgians, their place was taken by white hunters whose guns were much more powerful than the spears of the Pygmies. As a result, many gorillas were destroyed. Gatti estimates that between 1919 and 1932 about 250 gorillas were killed in Kivu alone. In most cases these were not killed for any scientific reason, but just for what was described as sportsmanship—in other words, so that the hunter could go back home and say that he had shot a gorilla; no doubt he took the skull and the skin back with him as evidence of his prowess.

It was in this area that Carl Akeley had persuaded the Belgians to establish the Albert National Park, which has since helped to maintain the gorilla in this area. Gatti eventually got permission from the Belgian Government to capture a gorilla for the Royal Museum of Natural History in Florence. At last his childhood wish of being able to hunt the animal was about to meet with success. He, too, had his imagination titillated by the fabulous stories concerning the gorillas, stories both real and fancied, such as that a strong man can be crushed like an eggshell with one squeeze of a gorilla's hand; that a gorilla could bend the barrels of guns double just as easily as a human could bend a toothpick in his fingers; that he would invade villages in herds and destroy the huts and the gardens. Gatti also heard wondrous tales of titanic fights between huge male goril-

160

las and snarling leopards or between some tremendous python and a huge male gorilla.

After progressing some distance along the shores of Lake Kivu and going some distance out from Bukavu, Gatti and his comrades camped one night and at sunrise were awakened by an elemental scream they presumed was a nearby gorilla's greeting to the rising sun. Gatti gives a picturesque, imaginative, and probably incorrect description of this occasion; he said, "Each morning, in just a few seconds before the first rays of the sun peep above the horizon, the gorilla comes out of his shelter, stands erect, alone, beating his breast with his huge man-like hands, and at the first glimpse of the sun in the east raises that terrible piteous cry."

The Africans in that area called him Ngagi, which means in their language "the one that ends the night." Some of the Pygmies probably believed that the sun actually rose in response to the call from Ngagi. Gatti gave graphic descriptions of the gorilla forest; walking through it, he said, "it is not so much like going for a walk as it was like fighting every inch through a solid wall made up of trees or bushes and all kinds of heavily matted vegetation." He thought perhaps the gorilla had chosen this type of habitat just to cut himself off, to put a kind of impassable barrier between himself and human beings. Entering that forest seemed to Gatti "like passing through a door into another world, a world that had existed intact and isolated since the dawn of life. I seemed to have stepped into some prehistoric epoch."

One of the Pygmy guides of the Gatti expedition carried with him as a kind of talisman a piece of skin which appeared to have come from a gorilla, but which was a vivid reddish color. The Pygmy said that the skin had come from a huge male gorilla that had been killed by his father some eighty years before, deep in the heart of the forest, where no white male had ever been and where the Pygmies had not dared to go for a number of years. According to the Pygmies, there existed in that remote part of Africa some families of very large, more powerful, and more savage gorillas than those that lived in the Kivu district. Presumably it is a red gorilla, but such an animal, dead or alive, has never yet found its way into the possession of a white man, and it is a variety or a species completely unknown to taxonomic science. The possibility that it actually exists is very small.

Gatti heard and smelled gorillas, but could not see them; a

161

number of days passed and he did not see any sign of the herd he was interested in. One morning, however, one of the Pygmies pointed out the spoor of a number of gorillas and it appeared to be fresh. Among the spoor the Pygmies pointed out some very large footprints, and they called the animals that made them Moami Ngagi, or "the King of the Gorillas."

Shortly after this, Gatti found himself in what appeared to be a sort of natural balcony looking down at a small clearing in the forest, and in the clearing were a number of gorillas—one of them, a very big female, was basking in the sun and had a youngster of about three years jumping on her, nipping her, and generally making a nuisance of himself until it became too much for the mother. She then gave it a smack which sent it rolling over and over along the ground. A little way away from them were two other females and a large male, both covered with long black hair. The male was obviously mature and had a little gray in the center of his back. They were sitting tearing up bamboo shoots and stripping away the outside leaves from stalks of wild celery. At a tree near them one animal was digging in the ground for roots. A little bit farther along a male and a female were sitting side by side, with their backs resting against a tree and their long arms dangling down; their heads were sunken onto their chests; they looked like a corpulent elderly human couple taking an afternoon nap. There was one other animal, a half-grown gorilla that was grasping two liana vines and swinging back and forth on them. After a little while an enormous male entered the clearing on all fours; giving the impression of being immensely strong and powerful, he advanced, apparently with the air of being the boss, and looked around to see if his group was all in order. A baby entered at that moment, probably remembered some kind of previous chastisement, and immediately ran to its mother.

Just as this happened, one of the Pygmies with Gatti mimicked the cry of a leopard and immediately all hell broke loose among the gorillas. The young animal and the five female gorillas immediately climbed up into the branch of a tree. The animal swinging on the vines climbed up to join them in the branches. The old female gorilla that had been napping also scrambled up a tree trunk and sat high on the branches of the tree. All three males jumped into the erect position and stood with their long arms dangling and swaying around their knees and with their legs widespread and turning their heads from side to side, looking for the leopard attack. Captain Gatti was very impressed by the appearance of Moami Ngagi, for he was the last one to enter the clearing, and he looked to be six and a half feet in height and to weigh at least 500 pounds. He was covered by exceptionally long thick black hair, but on the back the hair was silverish. He beat his chest with one hand, making a loud, booming sound. The Pygmies repeated the leopard cry, and the large gorilla closed his hand around a tree, which he snapped across. Picking up the top portion of the tree and waving it, he advanced a few steps in the erect position, and uttered a howl, which might have been a warning or a threat. Whatever the nature of the call, the females and the young animal dropped immediately out of the trees and dashed into the jungle. As soon as they had gone, the big gorilla bit the trunk of the tree with his enormous canines and tore it up and then threw it away. Then, going down on all fours again, he slowly strolled away as if nobody could harm him.

Kasciula, one of the Pygmies, had been terribly anxious for Gatti to kill this particular animal. Gatti, after Moami Ngagi had gone, asked him why he was so angry about this animal. Kasciula claimed that a few years ago six of his subjects, of whom one was his own son, were filing along a path in the forest when Moami Ngagi suddenly jumped on them. They had not given him any provocation for this at all. The gorilla had seized Kasciula's son and another Pygmy and crushed them to death on his breast. The other Pygmies took off at high speed and with great terror. When Moami dropped the two Pygmies he had killed, he went after the fleeing group. The Pygmies apparently have an interesting hunting practice when they are chased by something. As they run along, they stick a spear in the ground at an angle with the point facing in the direction from which they have just come and from which their pursuer will undoubtedly be following them. The spear is usually concealed by the heavy grass on either side of the track. On this occasion Moami Ngagi was impaled by a spear that he ran into.

He stopped, according to their accounts, and howled furiously and tried to pull out the spear, which was difficult for him because the type of spear the Pygmies use has two curved hooks which embed themselves in the muscle. Calling out with pain and probably also with rage, the gorilla went off into the jungle, and he was not seen again for many months. One day he reappeared and seemed to be fully recovered.

Gatti found all this very hard to believe, but he did admit that when eventually he shot Moami Ngagi, he found a black spot on the belly which was devoid of hair and was obviously a scar in the place where the Pygmies had said a spear had penetrated him a few years before. The dead animal, on measurement, proved to be 5 feet 6 inches high and 4 feet 2 inches around the chest. He had an armspread of 7 feet 2 inches.

In 1929 the African Anatomical Expedition of Columbia University and the American Museum of Natural History left New York for Africa after having received official permits from the Belgian government to kill adult mountain gorillas in a region outside the Albert National Park in the Congo. The expedition was initiated by Professor Dudley J. Morton, associate professor of anatomy in the College of Physicians and Surgeons, who was interested in the structure of the feet, particularly in their evolution, and thought this might shed light on some of the foot problems that affected humans. He had dissected the feet of most primates, but had not been able to get at the foot of the gorilla. For a number of other anatomical reasons it was thought desirable to obtain two gorillas for the American Museum of Natural History and also to provide additional material for the Columbia University scientists. The associate curator of the department of Comparative Anatomy of the American Museum of Natural History, Henry C. Raven, was the leader of the expedition. Raven was also a lecturer in Zoology at Columbia University and well known in museum circles for having led zoological expeditions into a number of countries, including Borneo, Celebes Island, Australia, Greenland, and parts of Africa.

When the party arrived in London, they went to see Emile de Marchienne, Baron de Cartier, who was the Belgian ambassador to the Court of St. James's in London at that time. He had formerly been Belgian ambassador to Washington and had been a friend of Carl Akeley, and it was his influence which enabled Akeley to send to King Albert the plan to establish the Albert National Park up in the mountains in the eastern Belgian Congo, where it would provide a sanctuary *in perpetuum* for gorillas and the other wildlife that lived there. The Baron de Cartier told the new expedition that he had been so affected by Carl Akeley's enthusiasm for the mountain gorilla that he had become very much of a conservationist himself, and he was unwilling to allow even one gorilla to be killed for scientific or any other purposes even though he did understand that the studies they proposed to make on the anatomy of the gorilla would be very valuable. He also said that even if permission were given to take two gorilla specimens, and they carried out some valuable scientific work and published it, wouldn't this lead other specialists to question some of their findings and claim the right to kill more gorillas to check on the points that they were doubtful about? Eventually the baron gave them letters of introduction so they could explain their mission to the proper authorities when they got to Brussels. On arrival there, they were met at the railway station by Dr. J. M. Derscherd, secretary of the Belgian Office for the Protection of Nature, who was particularly interested in the gorilla since he had been a companion to Akeley in some of the latter's field expeditions. Dr. Derscherd helped the expedition get the permissions it needed, but they also needed to visit the French Congo to get a specimen of lowland gorilla. So they went to Paris. There they ran into a lot of opposition because the French government had been very resentful of the New York Zoological Society's protests about the selling of baby gorillas by animal dealers in France. Eventually they got an introduction to M. Antonetti, who was governor-general of the French Congo, and to M. Marchand, governor-general of the French Cameroons, and it was there that they eventually got their lowland gorilla specimens.

The expedition set up its camp on the southern end of Lake Kivu at an altitude of 7,000 feet; the camp faced eastward over the cultivated country, which extended between them and the lake. They could see the volcanoes to the north of the lake, and just behind their tents the forest began. As soon as the camp was set up, Raven made excursions every day into the forest, accompanied by two or three of his African guides. He started off with Bantu hunters, but found them not very good, and he finally managed to get some Batwa Pygmies, who are professional hunters, to help him, and they turned out to be much more successful. As with all the other people who hunted gorillas, he had several experiences of being rushed by large male gorillas.

Raven shot a 460-pound gorilla, black in color with a silvery-gray back and exactly what the expedition had been hoping to obtain. The big problem now was to get the animal down some considerable distance to the base camp, where all the material for injecting the blood vessels with formalin, a preservative, was to be found. The first stage was for the Pygmies to make a litter of small trees onto which the gorilla was lashed. Then a group of them set off to carry him down the mountain to be embalmed. They had to widen the trail down the mountain to about twelve feet for twelve miles, and Raven sought aid from the surrounding village to get the animal down. It actually took them two days to get the animal back to the base camp. Nevertheless, they were able to get good preservation, at least for anatomical dissection. Dr. W. K. Gregory describes the arrival of the gorilla into the camp:

> We heard the tumult and the chanting in the distance and soon the porters struggled up the slope to our camp, while the chants grew louder and fiercer. We could see the gigantic ape man in his white funeral wrappings, his immense abdomen swelling high above his mighty black chest. Finally as the bier came opposite our main tent, the toiling pallbearers raised the chant to a climax. With a mighty heave they raised the bier above their heads and then, taking a step backward, they let it down on the ground. No wonder the whole neighborhood was excited and we most of all.

So they had obtained one of the two gorillas they were permitted to take from the Belgian Congo. The second one they got near the base camp itself. A group of gorillas could be heard around the camp one night, and early in the morning Raven came out. A large male showed itself, but Raven didn't shoot at it because he already had a large male and was now trying to get a big female. Behind the male there was an animal with a broad face which he thought was a large female, and he took a quick shot at the face. There had been a loud bark from one of the gorillas, and they all rushed away. Later in the morning when the humans moved in to see what had happened, they found that Raven had killed the gorilla at which he had aimed, but it was not a female, as he had hoped, but another male. Also, the shot had actually missed the face, and the bullet had gone downward through the neck and chest. When they opened the animal up, they found that a number of large blood vessels which Raven had depended on to carry the preservative all over the body had been torn by the bullet. A great deal of time was spent trying to repair these vessels and to get the formalin into them so that good preservation could be obtained.

The two eastern gorillas were shipped off to America, and the party then set off for the French Congo or to some other place in West Africa where the lowland gorilla could be found. They had received permission from the French to obtain three gorillas from that area. They finally decided to try to get them from the Cameroons and paid an official visit to Governor-General Marchand in the town of Yaoundé to obtain their hunting permits. They decided in the first instance that they would try the country around the village of Ozoum, which was about fifteen miles northwest of Yaoundé. This was the West African country where Paul Du Chaillu had hunted his gorillas.

The gorillas in some parts of the Cameroons were a threat to the food and economy of the Africans. For example, in one place the party found that gorillas had raided a banana plantation and had destroyed most of the banana plants by tearing them down and breaking open the stems to get at the pith, of which they are very fond. Because of this, the Africans had very little sympathy for gorillas. On one occasion while the expedition was in the area, the screams of a young gorilla near one of the plantations were heard by the Africans, who went running down with their spears at the ready; on arrival, they saw that a young gorilla had been caught in a snare and a big male gorilla was trying to pull him out. The men made a loud noise and frightened off the male. Then they rushed up to the young one and cut off its hands with their machetes and killed it. Raven was very concerned about this, but was able to secure the head for his collection of gorilla material to take back to the United States. When the African chiefs in the area heard that Raven and his colleagues were interested in shooting a couple of gorillas, they were only too eager to organize a great drive and slaughter of gorillas which would have been done largely in revenge for what the gorillas had done to their plantations. But Raven refused to authorize this, for he could see no reason for such senseless slaughter of gorillas. As far as the Africans were concerned, it was not senseless because it would reduce the number that could damage their plantations. Their method of capturing gorillas was to spread nets in a number of areas and then form a large group of beaters who would drive the gorillas in the direction where the nets were set up. As soon as the gorillas were caught in the nets, the natives would rush in, spear them, and hack them to death while they were thrashing around in the nets trying to untangle themselves. The Africans were fond of gorilla meat, and W. K. Gregory made the following comment about their eating gorillas: "The eating of men being out of date, the natives find the gorilla a good substitute, and a certain white hunter 'Duala' told us that he had often been employed by the government to kill elephants and gorillas as food for the blacks that work on the railroad. From Raven's experience in this and other places in the Interior of the French Cameroon, he concluded that perhaps ten times as many gorillas are killed by the natives themselves as are killed by white men."

The party found a number of people who had gorillas in captivity in this area. In a town called Mbalmayo there was a Greek merchant who kept a young gorilla as a pet. He called this animal Alfred, and on market days the baby gorilla hung around the front of the shop and romped with a small son of one of the employees. The Africans who came to town to sell their produce found this gorilla provided some great entertainment, particularly as he behaved much more like a chimpanzee than a gorilla. He was dressed in clothes and was expert in hanging onto a post with his feet and one hand and then drinking from a can of milk held by the other hand. He was not an especially neat drinker and spilled a good deal of milk on his jumpers. If he got excited or if he found a special interest in something, he would beat his chest just as the adult gorillas in the wild did. There was another person in this town who had a baby female gorilla that was about eighteen months old. A parrot and a mongoose were also pets in this house. The small gorilla had the freedom of the house and the garden. For most of the day she would sit cross-legged on the ground and manipulate things with her hands, including the household kitten. She was very deliberate in everything she did and was expert and precise in her movements. Sometimes when she was eating a cracker, the mongoose would come up and try to grab it out of her hand or even from her mouth if she had it sticking out of her mouth. When he did this, she would turn her head away and put her elbow in his face; usually he had to make several sudden dashes from different directions before he was able to get a piece of cracker from her. Sometimes the family would tie a bunch of bananas from a horizontal support in the shed to see what the gorilla would do to get it. Her technique was simply to climb up the supporting vertical post, hang under the horizontal beam, and crawl along that until she came to the vertical wire. Then she would hang on by her feet and one hand, reaching down with the other arm to hug the bunch of bananas to her body. Sometimes she would simply break off a banana with a free hand and hold it up to her mouth, bite into it, and strip the skin off with her mouth. She was also keen on sugarcane and held it usually in one hand, eating it very much as a human child would.

In this young gorilla it was possible to see a differentiation between functioning of the hand and the foot. In the process of evolution this differentiation was complete in the human, but it was obvious that such differentiation was already under way in this young gorilla—the hands were primarily manipulative even though the knuckles were used to support the forepart of the body when the animal was walking. The feet, however, although used for locomotion, were not used for manipulative purposes at all, but they were used to grasp things. It may be that the habit of walking on the knuckles in both the chimpanzee and the gorilla is a method of preserving the sensitivity of the fingers and the sense of touch so that this would permit the animal to have a more delicate manipulative ability. In the orangutan, which walks on the flat of its hand, this is not so important because the animal is primarily an arboreal animal and comes down to the ground relatively rarely. This is not so, of course, in the case of the gorilla and the chimpanzee, which are primarily terrestrial animals and which spend a good deal of time walking on their knuckles. It would obviously be much better for them, if the hand is to be used for manipulation, to develop the thick supporting pads of flesh on the knuckles rather than on the flat of the hand or fingers.

There was a Reverend Johnson, a missionary in this area, who owned a gorilla called Bushman. The animal was three years old and, at that time, in vigorous health. He also had a smaller gorilla which was still a baby, but was much less rough and aggressive than the older animal. There was a black male nurse who looked after the young gorillas, and they would rush after him, grabbing him around his legs and then riding along as he moved. This made walking for the male nurse a laborious and clumsy activity. This tendency to clutch something seems to be built into the behavior of both young gorillas and chimpanzees. It probably is of value since in the wild it helps to keep them attached to their mothers and there is much less chance of their being separated from them. When Bushman and the other gorillas played with each other, they all seemed to recognize the fact that they were of the same breed, and they were not frightened of each other. The largest one, Bushman, was more aggressive than any of the others. He would grab the little one by the hind legs and drag her around. The little one would on occasion try to elbow him away and, if he persisted, would attempt to bite him. When she went for shelter under a bench, he would follow and be very rough with her; finally she would retreat and sit in the sunshine. He did not persist in following up these aggressions on the young one for too long; his mind was soon easily diverted to another idea, and he would try to climb over walls or push himself through holes in the wall and indulge in other playful antics.

After the expedition had been in this area for some time, its leaders decided that the rest of the party would go home and that Raven would go farther up-country, probably near the border of Gabon, in the hopes of getting two gorillas he was after. The others took a train from Yaoundé and took the little gorilla they had tamed, put her in a baggage car, and not surprisingly she screamed her head off there with a babylike cry; and even oranges, bananas, and sugarcane would not quiet her. At every station they could hear her screaming. She would stop only when they put their fingers through the spaces in the wire netting to give her something to clutch. It is perfectly obvious why she was screaming. Gorillas at that age are not that different from humans. To be put in a frightening situation in a cage in a baggage car and separated from humans that she had got to know would certainly be a devastating experience for her.

After the main party left, Raven and the Africans made an expedition to a little village called Tjambolo. There the gorillas were so close that they could often be heard at night. On one occasion Raven said he could detect their odor. One morning the Africans told him that the gorillas must have slept in very close proximity to their huts because they heard them talking and that the gorillas were probably feeding nearby. They went down to look for them in an area which was largely swampy and got into a plantation full of plantains and bananas. They were close enough to the gorillas even to hear their stomachs rumble. On several occasions they could actually see the vegetation move. Raven managed to get within forty-five feet of a gorilla but saw only an arm. Once he got a good view of a whole gorilla, but the animal did not remain still enough to enable him to shoot it.

Raven has described his many fruitless attempts to track and follow the gorillas through the forest in Gabon. A tracking party would be led perhaps only by what appeared to be the faint mark of a knuckle, a broken twig, or bit of the aframomum plant which they had chewed up and discarded (this is one of the gorilla's most common foods). On some occasions a tracking team would stray some distance off the trail when they would hear a branch break or some vocalization by a gorilla which would guide them back onto the trail. The gorillas "moved about in the forest as quietly as a cat walking across a carpeted floor." Sometimes the party would be close enough to see vegetation moving and then catch sight of a group of two or three or more gorillas. Then suddenly they would move off and it would be impossible to follow them. Sometimes some members of a group would move off, and one animal would remain behind making some kind of a noise and trackers would be under the impression that the whole group was still with them. Then suddenly the one that was left behind would steal off softly, and there would be no way of finding out where they had gone.

Once Raven came face to face with a gorilla and, raising his rifle, realized it was not the size or type of gorilla that he wanted. He lowered his rifle, and the animal turned and made off. Once it got out of sight it beat its chest, and this inspired other gorillas, presumably of its own band, which were in the neighborhood to make a terrific roar and stamp about in the undergrowth. Following that they all quietly crept away and subsequent searching showed no gorillas at all in the area. Perhaps, in this case, the beating of the chest had conveyed some kind of message to the other gorillas, or maybe it was just a coincidence.

The breast of one of Raven's porters was very badly scarred, and he also had some deep scars on his back, shoulder, and arm. Raven asked him how he received the wounds, and the porter told him they had been made by a gorilla a few years before. The story was that he had been out in the forest with some other Africans gathering sap from rubber trees. He got up one day at dawn to hunt some monkeys with his crossbow, which he used to shoot a little poison dart. He saw monkeys in some trees and went to shoot one when he suddenly found a large male gorilla on the ground in front of him. The gorilla rushed at him and stopped and then turned around and ran off a little way, stopped, turned around, and ran at him again. At this point the porter's nerve gave out and he started to run; the gorilla grabbed at his rear end and missed, but caught him by the ankle. He hit the animal on the hand and also on the face with his crossbow. The gorilla stopped and retreated for a few yards, and at that point a female gorilla descended from a tree in the vicinity and the male made another rush. This time the porter hit it in the face with his fist and at the same time called for his friends to bring him a knife, but there was no one within hearing. The gorilla again retreated, this time only a few feet, and then made another rush, this time grabbing the human by the ears and trying to bite him in the face, whereupon he lowered his head. The gorilla, which was holding his right hand, then tried to bite it, but the man managed to yank it out of the gorilla's mouth before it could sink its teeth in, but the canine did inflict a bad wound on the back of his wrist. The animal still held him, however, and bit him on the chest, shoulder, and arm. He continued to strike at it, and eventually it released him and he ran away. Raven believed this story, even though this type of experience is rare. There is no doubt that occasionally gorillas do have fights with the Africans, particularly since they are often competitors for food.

On another occasion a woman who was supposed to go down to a planatation to gather some plantains and bananas wouldn't go because the gorillas were down there. The men told her that if she made enough noise, the gorillas would run away. She did this, but it led a large gorilla to make a rush at her. This time, instead of stopping, he simply ran past her, but as he ran past her he gave her a backhanded slap that was sufficient to knock her down and almost tore one breast off.

Raven makes an interesting statement about the relationship between gorillas and Africans:

> For centuries past the gorillas and natives have been competitors. As the native population increased, new villages would be formed and more clearings made. Then epidemics would occur, killing off great numbers of natives, and their gardens would be neglected to run into second growth. The gorillas, with a constitution so nearly like that of man that they can find more food in human plantations than in the virgin forest, would move into these deserted clearings. There with an abundance of food they thrived and congregated to such an extent that eventually if only a few natives remained, they were actually driven out because of their inability to protect their crops against the gorillas. But with the advent of the white man's government, with the distribution of firearms among the natives, preventive medicine and the treatment of epidemic and infective diseases, man has the upper hand at present in this age-long struggle.

There is no doubt that some African tribes feel a relationship to the gorillas. In *The American Anthropologist* of 1911 there was quoted a story of creation which was believed by the Bulu tribe in the Cameroons. The story went like this

God had five children; the gorilla, the chimpanzee, the elephant, the pygmy and man. Each was given fire, seeds, and tools, and told to go out alone into the world and settle down somewhere. The gorilla went first, and as he traveled along a forest path he saw some delicious red fruit. After he had eaten his fill and returned to the path, he found that his fire had gone out. So he stayed in the forest living on fruit. The chimpanzee left home, next, and he too became hungry as he passed a tall tree in fruit. He climbed up, and when he returned the fire had gone out. So the chimpanzee, like the gorilla, went to live in the forest. The same fate befell the elephant.The pygmy went much farther than the others and finally cleared away some underbrush, planted a seed and built a small hut; he did not cut down the tall trees, but he kept the fire going and learned the ways of the forest. Then at last man ventured forth and traveled very far. He cleared a large garden and cut down the trees, and built himself a large house. He burned the brush and planted his seeds and lived there until the harvest was ripe. After a time God went out to see how his children had fared. He found the gorilla, the chimpanzee and the elephant in the forest, living on fruit. "So," said God, "you can never again stand before man, but must ever flee from him." Then God found the pygmy under the trees and said, "So you will always live in the forest, but no place will be your fixed abode." Then finally he came to man and saw his house and garden and he said, "So your possessions will remain always."

After he had shot and preserved his first lowland gorilla, Raven got seriously ill and after a long period in which his life was in danger he eventually recovered, but the doctor who treated him tried to persuade him to go back to the United States. He was very anxious, however, to get the second of the two gorillas that he wanted and he decided to let the Africans do the hunting and bring to him what they caught. According to Raven, Dr. Lehman, the doctor who had looked after him, told him that he had contracted sleeping sickness, two types of malaria, hookworm, and ascariasis. The last disease is caused by an infection in the intestine from the roundworm called ascaris, but Raven, in addition, had a number of other intestinal parasites. By remaining another nine months in the forest and letting the Africans do the hunting, Raven was able eventually to obtain two more adult male gorillas, which he embalmed. He had also obtained some chimpanzees and got some skeletons of both gorillas and chimpanzees from the Africans. He finally secured another adult male gorilla before he left.

With all his material he eventually set sail for the United States on January 5, 1931. In a postscript to the book about the Raven expedition Dr. W. K. Gregory was at pains to point out that Columbia University and the American Museum of Natural History stood, as he put it, "for the protection and conservation of the gorilla." He drew attention to the fact that Raven had not attempted to exceed the quota of animals he had been officially permitted to obtain. Furthermore, he had had an opportunity to see how the demand for gorilla skulls by scientists at that time had affected the gorilla population in Africa. This demand was so great that both white men and Africans had a very profitable business killing gorillas and selling their skulls. This had caused a serious decrease in the gorilla population. He believed that in many areas the gorilla was actually in the process of rapid extermination.

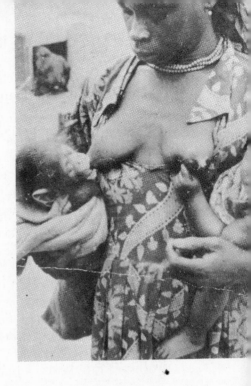

Wet nurse for great apes. Gorilla baby suckles at breast of African woman. (Courtesy of Mae Noell.)

7

Dr. Yerkes Sets His Scientific Sights on Wild Gorillas

WHEN Dr. Robert Yerkes was establishing the Yerkes Laboratories for Primate Biology in Orange Park, Florida, in 1929 and 1930, he sent Dr. Henry Nissen off to study chimpanzees in the wild in Guinea and West Africa and arranged also for Dr. Harold C. Bingham to go to Africa to look at the mountain gorilla. Dr. Yerkes believed that the knowledge of great apes would be incomplete unless they were studied in the wild as well as in captivity.

Bingham went to Africa in the year 1929-1930 as part of a joint expedition of Yale University and the Carnegie Institute of Washington for the psychobiological study of mountain gorillas. The study was to be carried out in the Albert National Park, in the Belgian Congo. Bingham had carried out some experimental studies of the gorilla Congo, which had been brought to the United States by Ben Burbridge.

In the spring of 1927 the Belgian ambassador to the United States and Dr. John C. Merriam, who was president of the Carnegie Institute of Washington, suggested to Dr. Bingham that he undertake a scientific expedition to the Belgian Congo, in the company of the ambassador, and that the main purpose of the expedition would be to make a survey of the primate resources in the Albert National Park and to offer some recommendations concerning the use of that region for the scientific study of its animals. Bingham was forced to decline the invitation, but two years later, in 1929, Dr. Yerkes' plans for setting up the beginnings of his primate laboratory had advanced so far that it seemed an appropriate time to undertake such an expedition. The idea was initiated by Dr. Yerkes, and Bingham agreed to go.

Although both Dr. and Mrs. Bingham went on this expedition, they were handicapped by having no previous experience in Africa or with any type of tropical life, and they lacked acquaintance with the mountain gorilla, the object of their search. Both, however, were very well trained in the methods of comparative psychology. They had a good general knowledge of anthropoid apes, at least as captive subjects, and they were intensely dedicated to learning more of the life of the gorilla in the wild.

The Binghams took a ship from Europe to Dar es Salaam and then a train and a boat via Kigoma and Kisenyi to the gorilla sanctuary located in the vicinity of Lake Kivu. Kisenyi is actually a port on the north end of Lake Kivu. Wild gorillas were within less than a day's walk from either of these port towns. The Binghams found many trails of gorillas and also heard them on a number of occasions. Once they followed the trail of a single gorilla group for 100 hours.

They obtained considerable information from this 100-hour trail. There were twenty or more animals in the group they followed, and the animals kept them continuously on the move for four or five days. It is important to remember that Bingham's interest was not to capture a gorilla or kill a gorilla for any museum or other purpose, but simply to observe the animals in the wild.

At the start of the trail the animals climbed the precipitous and slippery walls of a watercourse and then spread out when they got to the top to enjoy the succulent vegetation they found there. Then they formed into a single file and moved off, leaving a conspicuous trail behind them. On the first part of their journey the gorillas climbed over the crests of hills and mountains. They meandered through ravines in which there were many large trees and in which the undergrowth was extremely dense and luxuriant. Each day the tracking party came across a set of nests that the animals had slept in the night before and were able to make a study of them. After the second night the animals had ranged individually for feeding purposes for a relatively short distance, and then their tracks all converged again at the foot of a large tree. From this tree there was a direct path that was very well defined, which suggested that it was frequently used by animals, and particularly by gorillas, traveling in this region of the mountains. This path led up to the first fairly acute slope of Mount Mikeno and then followed along a nearby crest. At this stage the gorillas ceased to make day nests, and Bingham thought that their travel was more direct and more hurried. On the third day the trail passed through dense nettles and a very complex meshwork made up of creepers which had a retarding effect on the progress of the humans and made following the trail difficult. Mrs. Bingham, in recalling the final twenty-four hours of this 100-hour trail, wrote:

Heavy rains had left the saddle region messy. The first day's climb up Mikeno brought to view a panorama of the nest group scene at the same altitude (9,500 feet) on Karisimbi. Here the ascent was steep; the two groups were separated by 100 feet; each was in a clump of brittle semicios, the one almost directly above the other. Adult and infant groups (that is parental and non-parental nests) were compared and numbers tallied with the preceeding night. Now the story totalled three days and three nights. Gorillas had fed, rested, slept; fed, trampled, slept; fed and rested, slept; now they were headed beyond the first slope of Mikeno. Out of the tangled growth of fresh celery which was coming through the rotting stalks of the old; on through a region where the ground was covered with cushions of moss, and the heather was hanging hoary-like, the guides trailed those gorillas. Where fresh grain, brittle stalks of celery and other plants are thick, gorillas leave a flattened trail easy to recognize for it is crushed down, as by caterpillar tanks. In the moss the trail is not so clear; but it is helped in a few places where there were steep descents on the other side of the 10,000 foot ridge. In some places those slippery paths seemed impossible. Down, down the beasts had gone, actually sliding in chutes when there were no tracks to be seen. There were few roots or bushes to hold to, and the earth or bare cliff showed where they had tobogganed to the bottom. In the ravine a sparse growth of celery had been found, and a rest in day nests had followed. Our guides did not stop for rests this time; and slope number two seemed to inspire them. The trail crossed a tiny stream, and followed a laborious course along the opposite ridge. The signs of the trail now read only two hours old. It was going west along the side of the ridge, but circled back in a sharp loop, 200 feet nearer the crest.

Turning abruptly to the right they travelled through another area of laurel, moss and heather at the same altitude as ridge number one (10,000 feet). Again a descending trail through a sparse growth of celery led to another jungle of fallen trees, and decaying vegetation, and growing plants so firmly matted with vines in places that even the heavy gorillas left trails that were not too easy to follow. When we were within 200 feet of the bottom of the ravine, the gorillas were seen peacefully gorging themselves. They sat in open places at the higher union of slopes number one and two. . . . A long lobelia was grasped in one or both hands, a firm bite was taken, and then as the arms pulled left the head turned right, to tear off the tough exterior before swallowing the juicy heart. For several minutes the animals were quiet. Each was seated in the opening partially cleared about itself as it pivoted in the day nest, extended the long arms, and broke off the fresh succulent food. About the large body of each there was a growing pile of telltale shreds marking the rims of nests as we had seen them so often along the 100 hour trail.

This area of two ridges had no signs of earlier bands like the Karisimbi side of the triangle. The trail led to no old nests, stamping grounds, or feeding places. On the Karisimbi slopes evidences of long occupancy were everywhere about. Is that region of Karisimbi a seasonal rendezvous, or permanent camp, and do the animals prefer it to Mikeno? One might suggest as an answer that it is preferred when the sun shines from the north side of the equator, but who can tell us whether the wanderers go still lower to the bamboos of Rueru when the mwanzi mtoto (young bamboos) are in season.

This description is a graphic account of the somewhat aimless and somewhat deliberate wanderings of a group of gorillas, and is one of the earliest detailed records of what a group of gorillas does when it is on a day-to-day march, and from this point of view alone, it was of considerable scientific importance in our understanding of the life and habits of the mountain gorilla.

Bingham established the identity of a dominant male leader of the group that had been followed, and called him Bwana, but at that stage he was not certain of Bwana's relationship with the other members of the group. He was the only adult in the group that showed excitement on occasions. It seemed that the cues for walking, resting, feeding, and other activities were being signaled by this leader, the old silver-back male. Bingham and his wife were able not only to record the way the gorillas moved about in their environment; they were able to study their feeding and sleeping behavior. They collected and identified a number of the plants that the animals subsisted on, which included the wild celery, the wild carrot, certain vinelike stems, and bamboo.

Water did not seem to be important to the gorillas. This has been observed in many field studies. Gorillas would cross a shallow stream if it were in their path, but they avoided running water as much as possible and would even, if they were evident, use stepping-stones to get across, a method used also by the chimpanzee. Gorillas do not come back to sleep in the same sleeping site and reuse the same nests. The animals find it simpler to sleep near the food supply than to keep returning to a sleeping site at night. The gorilla philosophy is that it is better to move on and to sleep where the food is, a philosophy still followed by some Australian aboriginals. Also they do not carry nest-building material any distance to their nests; they build a nest where the materials are already situated, a real economy of effort. A lot of nests are on the ground, but they can also be built up in trees and range from about ten to sixty feet in height.

On one occasion they came across a group of gorillas: "In drawing nearer, we were discovered. That brought eight or nine gorillas into the foreground where they sat as if in a meeting. Among them were four gray-backs who took turns standing up to second the motion. They stood, barked, clapped their chests, and sat down again. This performance was imitated by two youngsters."

One day when the party was following a gorilla trail, they sensed an air of excitement among the animals ahead of them. Bingham, feeling uncertain, took his rifle from the gun bearer and they continued trailing. Mrs. Bingham's notes contained this entry:

Following the guide's lead down the slope, we made a wide circle from the place where we last heard the gorillas. As we climbed the next ridge, the guide paused to look at a freshly uprooted lobelia and to say something about ngagi ngini (gorillas again). . . . We had gone only a short way through vegetation higher than our heads when a gorilla screamed. The guide stopped, whirled about, and fled. Mr. Bingham called for me to come back, which I turned to do, but I had only taken a step or two when he commanded, "Get out of the way." I pushed into the carrots at the left of the trail while he ran forward and fired.

Gorilla feces in the path showed that the animal had advanced within fifteen feet of the spot where Dr. Bingham stood when he fired. The gorilla turned about and fell in the trail forty paces distant. It had run on after it had been shot and finally dropped dead in its tracks. Mrs. Bingham was within three paces of where the animal was shot. Bingham claimed that he had saved his wife by that shot. We do not know whether the gorilla was only indulging in one of its bluffing runs. No one, however, can blame Dr. Bingham for not taking any chances. Bingham's gorilla was 5 feet 7 ¼ inches in height and his arm reach was 9 feet ¾ inches. Quite a big and intimidating animal! He was full grown, of course, and had all his permanent teeth. There was no way of really weighing him in that isolated area.

During the 1930's some interesting observations on gorillas were made by Charles Pitman, a game warden in Guinea. He also found gorilla nests in the trees, confirming that the animals still climbed trees to some extent, although they are mainly terrestrial animals. Pitman made an expedition with the governor of Uganda to see gorillas in the wild. A group of people followed a group of apes for some distance, watched them, and then decided that they had seen enough and returned to their camp. The gorillas, however, not to be outdone by the humans, decided that they wanted to do a little observing of their own. As the party withdrew, the gorillas turned around and followed them. As they approached the governor's party, a big male gorilla barked at the humans. Pitman had this comment to make: "For a game warden it was an unenviable position. On one hand the sacred person of the Governor, on the other hand the almost sacred and strictly protected gorilla. . . ." But Pitman, the governor, and the gorilla maintained their sangfroid, and everything ended satisfactorily. The entertaining part of the story is that the gorillas returned the compliment the governor had paid them by going into the jungle to see them.

8

Learning About Gorillas Through Binoculars

THERE was very little gorilla observation in the 1940's because of the outbreak of the Second World War. It was not until the 1950's that expeditions began to be made again into gorilla country, especially into the domain of the mountain gorillas in the Virunga volcanoes of Albert National Park. Spearheading these scientific studies was Walter Baumgartel, who was fascinated by these giant apes and wanted to encourage investigation.

So in 1954 Baumgartel came to a remote corner of Western Uganda to the region known as Kisoro. He knew that this was part of the gorilla sanctuary, the Albert National Park, and he knew that the gorillas were there, but he also knew that only a handful of people had ever seen them and wondered how to remedy that. He made the acquaintance of a local African called Old Reuben, who had been born and raised on the edge of gorilla land and was familiar with the forest and mountains of that area. He had requently seen the gorillas, and he took Baumgartel out to try to find them. Baumgartel's objective was to map the gorilla trails and then invite the sceintific community to come to study them.

It should be mentioned that the Virunga volcanoes form a chain across the floor of the Western Rift Valley, 100 miles south of the equator. Two of the volcanoes are Nyiragongo and Nyanlagiro, which are situated at the west and were active in the 1950's. In addition there are the three volcanoes mentioned earlier, Karisimbi, Mikeno, and Visoke. They form the central group, and at the east there are the extinct volcanoes known as Sabinio, Mgahinga, and Muhavura. The northern slopes of these are in Uganda, and they form a gorilla sanctuary. Muhavura is 13,547 feet high. Reuben and Baumgartel trekked all over the eastern volcanoes and came close to the gorillas; they heard branches crackling and could smell the animals, but for a time did not see them. On one occasion Reuben was thirty yards ahead of Baumgartel; suddenly to the right they heard a crashing in the forest. The two of them stopped dead in their tracks, and after a few seconds a gorilla family strolled past the bamboo thicket in a very leisurely fashion and onto the path between them. The male seemed to take a very unhappy view of the situation when he saw Reuben, and he jumped over a narrow water channel and disappeared into the forest to the left of the trail. His wife, not really knowing what was going on, obediently followed him, but the youngster hesitated and apparently, being inquisitive, wanted to know what was going on. He looked first at Reuben and then at Baumgartel. They were possibly the first humans—the latter was certainly the first white man—it had seen. Then it suddenly realized it had been left alone with these two strange human characters for company. Its courage failed and, forgetting that its parents had crossed the road and gone into the forest, turned back and ran into the forest where they had all come from originally. The mother, realizing the child was not with them, returned to the trail just in time to see the youngster disappear into the thicket on the other side of the road. She immediately chased over the road to him, grabbed him by the hand, slapped his bottom, and dragged him over the path into the forest, where the father was barking impatiently. This was about as human a scene as one could imagine.

Baumgartel wondered if he could entice gorillas to a particular spot by spreading food in their paths. He tried bananas, sugarcane, sweet potatoes, maize cobs; but he found the gorillas were very fussy as far as feeding was concerned. They didn't like his newfangled dishes. They were prepared to settle for juicy bamboo shoots, wild celery, parsley, nettles, bark, roots, giant lobelia, and a plant called red-hot poker. Eventually, however, he found that salt licks had some attraction for them.

In 1955 Baumgartel was able to acquire the Traveler's Rest Hotel in Kisoro, Uganda, which is situated at the foot of the Virunga volcanoes. He had hoped to establish the place as a tourist attraction and as a base for scientific studies. Baumgartel had contact with Dr. Raymond Dart, the discoverer of the prehuman *Australopithecus,* and Dr. Louis S. B. Leakey of the Witwatersrand University and Coryndon Museum. Through his interest these two scientists made funds available for other investigators to study gorillas in that area. By this time Baumgartel needed an assistant and appealed to Dr. Leakey. Leakey sent his secretary, Miss Rosalie Osborn. Rosalie was only a girl of twenty-two at that time. She had considerable courage and on many occasions camped out overnight accompanied only by Reuben and his trackers. She was the first young girl to do this sort of thing. She was not a scientist, but she had a natural gift for observing and kept a very methodical diary of her gorilla observations. She was there for four months. After she left, Jill Donisthorpe, a trained zoologist, took her place and spent eight months studying the gorillas. She published her investigations in the *South African Journal of Science*. Rosalie Osborn published her findings in a symposium of the Zoological Society of London called *The Primates*. Vegetation is a big hindrance to the investigator in studying gorillas. In this area they lived mainly below 11,000 feet, and the vegetation there was particularly dense, so that an observer could be within thirty feet of a gorilla and not see it. The animals do not remain in any particular fixed territory, and they could wander any distance from 100 yards to three miles a day. Rosalie ascended the volcano Muhavura on a number of occasions in the company of African guides and usually came back about 7:00 P.M., before it was dark, but had no success at contacting gorillas using this technique. Finally she and the Africans built bamboo huts at 10,000 feet on the saddle between Muhavura and Mgahinga, at the very top of the zone where the bamboos grew. The procedure was to set off after breakfast, never to be armed, and to go with Reuben, who was an expert guide of the area, and one or two porters. The group saw considerable evidence of gorillas in the form of nests, food remains, and dung. They soon learned how to follow the tracks and Rosalie saw four gorilla troupes on Muhavura. One of these was composed of only three animals, and the largest had five. Rosalie and Miss Donisthorpe made a very good list of foods that the gorillas ate, and neither of them saw any evidence that the animals ate meat or even any insects. Nearly all the gorillas seen were quadrupedal, that is, they moved on all fours, and only once, and then for only a few seconds, was a gorilla seen to stand up on his legs. She noted that they had a variety of vocalizations. They would grunt when they fed, making noises like *huh, huh, huh*—sometimes as many as thirty of these in succession. There were barks often with three notes in each bark, and screams.

In 1957 Dr. Kinji Imanishi and Dr. Junishiro Itani visited Baumgartel and observed gorillas in that area. The Japanese created quite a sensation among the local people when they arrived, as they were the first Japanese ever seen in the area. The Japanese were successful in encountering gorillas three days in succession, and on the fourth, when they were climbing Mount Muhavura, they came across another group just below the summit of the mountain. The two scientists had a novelist with them, Taisuke Fujishima. He admitted that the gorillas had scared him stiff, and when a silver-back male barked at him, he ran away. By the time they left, Baumgartel said that a Japanese team under their leadership would always be welcome.

The Japan Monkey Center, which had sponsored the visit of Imanishi and Itani, sponsored a second Japanese gorilla expedition in April, 1959. This second expedition had two investigators, Maseo Kawai and Hiroki Mizuhara. They made many sorties into the gorilla areas and heard drumming by the gorillas at 4:30 in the morning. This, they said, was the time that the gorilla woke up, but after that they remained in their nests and lazed around until about 7:00 A.M., although they were observed in some cases to stay in their nests later than that. It was observed that gorillas would rest on two separate occasions during their daily wanderings, before noon and also in the afternoon. The rests in the morning were of short duration, but the afternoon rest period usually involved making a small nest. The distance that they go in a day is inversely related to the quantity of the food in the area. The more abundant the food, the slower they move. Although they do considerable wandering around, the distance traveled in a straight line is rarely more than a kilometer.

The gorilla in this area depends on the bamboo shoots as its main food. The bamboo shoot appears in June and July, but by August the shoots have matured and are too tough for the animals to eat. During the succulent bamboo season the gorillas depend on it for about 90 percent of their food supply.

Japanese investigators were able to approach in some cases within ten to twenty meters of gorillas in thick forests. This suggests that their auditory function, their ability to see, and also their ability to smell would compare with the degree of development of these senses in man. The gorillas, the Japanese observed, rarely climbed a tree, but ate thirty-four different species of higher plants of which at least eight species had a bitter taste, and one fungus.

The Japanese investigators made some interesting observations on nests. For instance, the baby gorillas sleep with their mothers in the same nests, but as they grow larger, the mother will accommodate the baby by building the nest into a gourdlike shape. In the larger part of that sleeps the mother gorilla, and in the smaller part the baby sleeps. When the baby gets too big for this arrangement, the nest separates into two parts, a large nest and a smaller nest, but the baby's nest remains very close to the mother's. The investigators were not able to observe whether the mother helped the baby build its own nest or not. Among the surviving anthropoid apes, the gibbon is the only one that does not make nests. Gorillas, chimpanzees, and orangs all do.

Baumgartel's tracker, Reuben, claimed that he once fought with a young female gorilla which tried to strangle him, but gorillas among themselves seem to avoid confrontations of force. In fact, one group of gorillas will consciously try to avoid running into another group of gorillas. This, of course, makes coexistence much easier. Unlike baboons and man, there is no display of greed among gorilla groups, a situation which some scientists believe represents a higher level of mind. Like other gorilla observers, the Japanese also recorded mock charges, but on one occasion a silver-back male with a malevolent expression charged them and knocked them down. The other tracker stood still and swung at the gorilla with his *panga*. The gorilla continued to rush him, however, and struck him on the arm, but then turned and ran off. Fortunately none of the four received any serious injury, although the situation could have proved fatal to one or more of them.

In 1958 Dr. G. Schaller planned with Dr. John T. Emlen, Jr., professor of zoology at the University of Wisconsin, an expedition into the eastern part of the Congo and Western Uganda. They wanted to study the range of the mountain gorilla and also to make continuous observations of the life history of this animal for about a year. The expedition, which was called the African Primate Expedition, left New York on February 1, 1959. It included Dr. Schaller, Dr. Emlen, and their wives. After the first part of the project had been completed, Dr. and Mrs. Emlen came back to America, and from August, 1959, to September, 1960, Dr. Schaller and his wife, Kay, devoted their entire waking hours to a study of the gorilla population in the Virunga volcano area. Schaller, working alone and unarmed, believed in the concept that the gorilla fundamentally is a shy and retiring animal and doesn't try to make trouble unless humans harass it. He noted that Bingham had had to shoot a gorilla to protect his wife and pointed out that Akeley, whom he described as the gorilla's best friend, had commented that, "I believe, however, that the white man who will allow a gorilla to get within ten feet of him without shooting is a plain darned fool."

Schaller, however, felt that there was no point in taking a rifle or revolver into the field with him. Firearms did not have a place in the type of study that he was planning, and no animal will attack without good reason and even then on rare occasions. He was prepared to give any animal that was charging the benefit of the doubt and assume that it was merely a bluff. If the charge were to be actually carried through, it would be probably too late to shoot the animal anyway, so Schaller reasoned there was no point in taking a gun. Schaller's wife, however, was not too happy about this plan, so he compromised and took to carrying a starting pistol that is used to fire blanks at track meets. At least he could fire it and make a noise. He never had to use this innocuous weapon.

At the base of Mount Muhavura is the little village of Kisoro. As we mentioned, Walter Baumgartel's Traveler's Rest Hotel is situated there. There were some shops there managed by East Indians until the dictator who now controls Uganda forced the Indians to leave the country. The shops originally carried all kinds of goods and cloth, kerosene lanterns and cooking utensils, etc. The road to the Congo passes through this village, although a new road has been built so that it is not necessary now to go through the mountains.

Schaller stayed at Baumgartel's Traveler's Rest and described it as being a small hotel composed of a main lodge surrounded by beds of flowers which were colorful and added a gay appearance. Three small, round huts were on one side of the main lodge. There was no running water at the Traveler's Rest—at least not when Schaller was there—and no electricity, and therefore dinner had to be served at night by lantern light. "It's the only place in Central Africa," Schaller pointed out, "which the average visitor can in one day enter gorilla country with a fair chance of seeing the animals." There are now other places where gorillas can be found close by. It is also possible to find gorillas close at hand near Bukavu.

One of the first things that Schaller confirmed about gorillas is that they do not like contact with water. One group of gorillas was seen to cross a small stream which was about eight feet wide and not more than two feet deep. The group first crossed by means of walking along a fallen log, and the next time they broke down three tree ferns so that the stems fell across the stream and bridged it. On the third occasion they jumped from the bank onto a rock in midstream which was just under the surface and then jumped from that onto the opposite bank. Schaller and Emlen found also an area where the nests of both gorillas and chimpanzees were present, though they were never fortunate enough to see chimpanzees and gorillas together. Their guides liked to tell stories and sometimes would sing a Swahili song to which they improvised choruses when something happened that was appropriate. What they added to the song fitted the situation. The song was unending, being improvised as the party went along. One of the guides wrote a translation of one of the verses:

> Karisimbi, Mikeno, and Sabinio are mountains wet and cold,
> Nyira gongo is a mountain of fire.
> They say that one can see gorillas on Mount Tshiaberimu.
> We're tired of searching for those gorillas because they exist
>   only in the imagination of the gods.
> Peace to our Mother, Peace to our House,
> Happiness to our House, those who were born Rutshuru.

In the Mount Tshiaberimu region Schaller found that gorillas were still hunted for food even though at that time, in 1959, they had been protected for more than twenty-five years. It is very difficult, however, to enforce protection rules in remote mountain villages, and Dr. Emlen found skulls of fourteen gorillas in different villages, which indicated that there had been quite a lot of killings. As a result of hunting the gorillas, many of the Africans get injured by the animals. The Kitsombiro Mission Hospital reported that they treated nine injuries caused by gorillas in as many years. In one attempt when the Africans surrounded a lone silver-back male in order to kill him, three of them were bitten by the animal before he was subdued. In the same area just before Schaller and Emlen arrived, a black-haired male had wandered into a wattle grove. Here the Africans surrounded him with spears, and the gorilla, provoked, chased an African, caught him by the knee, and stripped a piece of muscle seven or eight inches long off his calf. Moralizing on this event, Schaller quotes George Bernard Shaw, who said, "When a man wants to murder a tiger, he calls it sport; when a tiger wants to murder him, he calls it ferocity."

Emlen and Schaller visited the town of Bukavu, in what was the Belgian Congo, now the Republic of Zaire, and described it as being a very beautiful town built on the shores of Lake Kivu. Some twenty-five miles north of Bukavu is situated the research headquarters station of IRSAC, the Institute for Scientific Study in Central Africa. At this institute studies are carried out in anthropology, protozoology, virology, botany, and a number of other scientific fields. Even today mountain gorillas can be seen only a short distance away from the IRSAC station.

Schaller was disturbed, as most of us are, by the method of collecting young gorillas. He mentioned one unscrupulous American dealer who shot down entire gorilla groups in order to obtain the infants. Part of the dealer's report read:

> On the other side of a small clearing a female was playing with a small baby. Everything seemed perfect for a good shot. The gorilla wasn't fifty yards away and was unconscious of our presence. But I couldn't help thinking of the other gorillas that we could hear, and couldn't see . . . but there was nothing for it, I had to shoot it. I took plenty of time and when I stopped shaking, I made a clean hit through the skull, killing her instantly. [The male rushed up.] I fired and hit him in the shoulder. He staggered for a second, but kept going. I fired again, and again, he staggered.

Schaller points out the infant that had been obtained by the slaughter of its parents lived only a few days. Although the collectors had planned to kill the parents to get their infant, they had made no provision for the care of the infant so obtained. For each of the eighty-five gorillas alive in the United States in 1964 (there are more now since a number have been born in captivity), at least five other gorillas had died while being captured or before they had reached their destination—probably a zoo. Schaller claimed that most zoos cared very little how their animals were obtained. Although there may have been some truth in that statement fifteen years ago, when it was made, we do believe that most zoos are concerned about this practice today. Charles Cordier, a Belgian dealer, has developed a more humane technique for getting gorillas. His method was to surround a whole group with nets. Usually about 100 Africans are employed to hold the nets, and the gorillas charge into them. Once they get tangled up, it is relatively easy to capture the young animals before the big ones get free. This is an expensive and a time-consuming method, but at least it is more humane. Any large animals captured by Cordier that do not break free are released. A small gorilla in 1964, when they were still legally traded, cost $5,000. It is our understanding that the gorillas which the Yerkes Laboratory purchased around 1964 were collected by Cordier. Cordier had led a rather hard life, but insisted on adhering to his principles. He had a number of pets, including many birds and a giant insect-eating pangolin, whose body was covered with overlapping scales, giving it a kind of fishlike appearance. There was a high stockade near Cordier's house which contained a large silver-back male gorilla called Uneedi, and in the house were two infant gorillas, Noel and Mugisi. An African boy was employed during the day to entertain Mugisi, who was two years old. During the night there was a watchman who had to baby-sit with the little gorilla. The small animal, Noel, slept in a crib in a bedroom of the Cordier house and would cling to Cordier's wife very tightly when strangers were present. The Cordiers were visited by Schaller and Emlen.

In his expeditions following his visit with Cordier, Schaller heard stories about the killing of gorillas. He was informed by one authority that in 1948 officials organized the killing of about sixty mountain gorillas so that eleven infants could be obtained and sent to zoos. Near the town of Utu a mining official told Schaller that he had shot nine gorillas simply for "sport." This was not an area where the animals are strictly protected, but that does not make it any less of a crime.

The gorillas' dislike of water again was demonstrated to Schaller at the Lugulu River. He observed that in one place the river was 250 feet across and was aware of gorillas inhabiting the northern bank, but none occurred on the southern bank. Farther up the river, where it was much narrower and where there were plenty of boulders that the gorillas could use for stepping-stones, Schaller found gorillas on the southern bank also.

Gorillas are probably better observers than man and can interpret what they see much more accurately, and in planning his observational technique, Schaller felt certain that if his movements were calm and if he was alone and showed no intent of harming the gorillas, they would soon realize they had nothing to fear from him. It is difficult for a man to get rid of all his arrogance and aggressiveness, so that he can approach an animal with humility and the feeling that he is in many ways inferior. But Schaller was able genuinely to do this. Control of one's actions is very important, for even casual movements can sometimes frighten the gorillas and make them suspicious about the intentions of the observer. The possession of a firearm is enough to alter the subconscious behavior of a human so that he may show an aggressiveness that he may not notice but which the animal is able to detect. If an observer meets a gorilla face to face, it is much more likely to attack him if he carries a gun than if he himself showed submissiveness.

Among most animals, certainly in the case of the dog and the rhesus monkey and man, a direct stare is a type of threat, and this is equally important to the gorilla. It is unwise to stare continuously at a group of gorillas, and it is even a mistake to train binoculars and cameras at them continuously. Schaller was very careful, when he was observing, not to look at the animals too long without averting his head. Although gorillas would accept one person nearby, they would often get excited if they saw two. So Schaller had to "park his wife" at home a great deal of the time, and the game warden, who went with him usually, had to go into hiding while Schaller himself watched the gorillas.

On one occasion the observer became the observed. Some gorillas were feeding on a steep slope about 100 yards above Schaller. He sat down near the base of a tree and identified the leader of the group. He called this animal Big Daddy because he had two silver areas in the hair on his back. The leader saw the observer and gave two grunts, whereupon a number of females and youngsters looked in the same direction and then went to his side in anticipation of danger. There was another male, whom Schaller called DJ, a striving executive type that had not yet reached the top—the second in command. Then there was a third male, even bigger than the leader, whom he called the Outsider, who roamed the periphery of the group. He was estimated to weigh between 450 and 480 pounds. Regarding weight, it is often claimed that male gorillas reach 600 pounds or more, and they often do in zoos, where they usually become obese. Of ten adult male mountain gorillas that Schaller heard of, which had been killed and subsequently weighed by the hunter, the heaviest animal had reached only 482 pounds, and the average was 375 pounds. To come back to the group, there was another silver-back male, whom Schaller called Split Nose because he had a cut on his left nostril. He was a young animal, and his back had only just developed a silver tinge. He made up for his relative youth by carrying out considerable vocalizing when he saw himself being observed.

DJ started the movement which ended with Schaller under observation. He left the place where he was resting and moved uphill. Then he slowly angled toward Schaller, always keeping a screen of vegetation between them. But he had to stand up every now and then to see where he was. The moment Schaller looked toward him he would look down and then sit without any motion for a time before continuing his stalking activities. He finally approached within thirty feet, roaring loudly and beating his chest. This demonstration completed, he peered out to see if it had any effect. Schaller was never able to get used to the roar of a silver-back male because it was usually very sudden and shattering. The effect that it had on him invariably was that it made him want to run, but at least he got satisfaction from the fact that the other gorillas in the group jumped when the roar started, too. At one point Schaller became nervous and began to climb the tree; he climbed to about ten feet. Then one of the females came to within seventy feet of him, sat on a stump, propped her chin on folded arms, and looked at him. Slowly the entire group came toward his tree. Schaller was a little worried about this behavior because it had never happened to him before. Females carrying infants came over to look at him, and two juvenile gorillas climbed another tree to get a better look at him through the branches and vines. One juvenile, estimated to be about four years old, climbed into a small tree only about fifteen feet from the tree Schaller was in. They both sat there, he recalls, nervously glancing at each other. One of the other gorillas, the only black-back male in the group, came to within ten feet of the tree and looked up at Schaller with his mouth open. And so it occurred that the observer was treed, completely surrounded by gorillas, who sat by the hour observing him. After a while the sunshine turned into hail, and the hail into a drizzle of rain, and finally the gorillas ceased their observations and set out to forage for food.

Schaller describes the typical gorilla day as follows: "Between six and seven they wake up, stretch their arms, yawn, reach over, break off pieces of wild celery or other material conveniently close, and eat breakfast. They also eat thistles and nettles, bamboo shoots, various fruits, sometimes the bark off several trees and a variety of leaves. Most of the plants they eat taste bitter to humans."

Schaller claimed he did not see gorillas eat any other animals

191

in the wild. He never saw them eat birds' eggs, insects, or mice even though he noticed there were opportunities where they could have done so. He saw them pass dead animals without handling the remains even though they were fresh. On another occasion they did not disturb any pigeon's nest that they passed by. But gorillas readily eat meat if it is offered them in captivity. In fact, the Yerkes gorillas have enjoyed cooked hamburgers. For breakfast in the wild the animals sit up in their nests, grab any food they can reach, move very little to get a piece of food if it is not convenient to them, and then go back to the nest and sit. As Schaller says, the only sounds you can hear are "snapping of the branches, the smacking of lips, and an occasional belch." The infants apparently watch what their mothers eat and learn from her what to eat and what not to eat, so that food habits are handed down from one generation to another.

The animals move around in a small area eating, and as the morning goes on, they eat less but wander about choosing tidbits here and there. They feed about two hours in the morning and worry very little about any other activity. Once they have finished their early-morning feeding, they move farther from each other and may actually move out of sight of one another. By about ten feeding has usually stopped. From midmorning the animals take a siesta that lasts until midafternoon. They lie around near the silver-back male, which is the leader, and soak up the sun. They lie on their backs, some on their stomachs, some on their sides, and others sit and lean their backs against the trunk of a tree. Sometimes they construct a nest in which to take their siesta. They can produce a serviceable nest in as short a time as five minutes. Schaller says of the silver backs,

> These males are also tolerant and gentle, and this is especially evident in the periods of rest. The females and the youngsters of the group generally seem to like their leader, not because he is dominant, but because they enjoy his company. Sometimes a female rested her hand in his silver saddle or leaned heavily against his side. As many as five youngsters occasionally congregated by the male, sitting by his legs or in his lap, climbing up onto his rump, and generally making a nuisance of themselves. The male ignored them completely, unless their behavior became too uninhibited, then a mere glance was sufficient to discipline them.

Not only do the animals sleep and doze and sit during the siesta period, but some of them also become occupied with grooming themselves. The juveniles are even more active self-groomers than the mature females. Infants, on the other hand, rarely ever groom since the mothers handle this chore. When the baby is still small, the mother will put it in her lap or over one arm and groom it by putting her fingers into its hair and pulling the hairs to one side. She seems to be especially careful of the grooming in the area of the anus, which bears a little tuft of white hair. The youngsters seem to dislike this procedure. Perhaps it is just being turned upside down that they dislike, but at any rate, they do a good deal of kicking while it is going on. A young gorilla is never physically chastised by its mother. Adults do not seem to groom each other unless one of them requests it. For example, one of two females sitting near each other may lean over and tap the other animal on the arm and then rise and turn her rear end toward the sitting female, who would then groom the area that had been indicated. When adults groom each other, they seem to confine their grooming primarily to the rear and back and to those other parts of the body that an animal cannot get at easily. Young gorillas, of course, use the siesta period as an extended period of play and an opportunity to explore the environment without fear of separation. They do a fair bit of climbing into the trees, and in this they differ markedly from the adult animals. In fact, the silver-back males may climb only a quarter as much as juveniles.

Gorillas spend about 80 percent to 90 percent of their waking hours on the ground, and when they climb, they climb in a very deliberate way. Gorilla climbing ability can be compared to that of a ten-year-old boy. They get hold of a branch, hold it while they get a safe foothold, and then pull themselves up onto the next foot and handhold. If they are not too sure whether or not the branch will support them, they sometimes pull it sharply to test it before they trust it with the weight of their body. They get down the trunk very easily by sliding down feet first, using the soles of their feet as brakes. There is usually very little aggressiveness within the group. Most of the quarreling is done by the females. They will sometimes sit and scream at each other and do a certain amount of wrestling and biting. When the silver-back male gets fed up with this, he goes over to the females and grunts at them, and they stop screaming immediately. They never seem to inflict any injuries during their quarreling. We saw a similar stabilizing effect with the Yerkes orangutan group in the zoo at Atlanta. There was quarreling among the females until the male was brought into the group, which then settled down under his influence, the females ceasing to quarrel in his presence.

Gorillas may rest up to three hours in the middle of the day. The silver-back male decides when the group ends the rest period, when they start to travel, and how far they will go. Once a group begins to move, the babies all rush to their mothers and climb on their backs. They tend to move in a single file when they are traveling in the forest, with the silver-back male at the head and often a young black-back male acting as a rear guard. They usually travel in the middle of the afternoon, and they do not go very far—half a mile would be a long journey.

Once they have reached their feeding grounds, eating is again the main activity, and they feed until dusk. Sometimes this is a rather leisurely period of feeding, with short periods of travel. As darkness falls, the animals become more and more relaxed in their movements and close in near the silver-back male. Then at 5:00 or 6:00 P.M. the leader of the group starts to break branches in order to build his nest; this is the signal for the other members of the group to do the same. So after a ten- or eleven-hour day they are all bedded down again for another night's sleep.

On one occasion on the slopes of Mount Mikeno the gorillas reacted to a group of ravens. Schaller, sitting and watching a group of gorillas and eating his lunch, saw the ravens flying high in the sky. Then he whistled, and to the birds it apparently suggested food because they began to circle around as they descended. As the shadows of birds passed over them, the gorillas ducked, the silver-back male gorilla jumped up and roared at the birds, and the females screamed. Some of the females looked at Schaller, and others looked at the ravens. The ravens, giving the impression they were enjoying this reaction, made a sort of dive-bombing swoop over them many times. This made the male get even more angry, and "the females milled about in utter confusion." The apes apparently didn't think of this as a game, although the ravens did. Eventually the birds landed near Schaller and helped him eat the rest of his lunch. They then flew down into the valley and returned after an hour and again flew over the gorillas. The male once more stood up and roared at the ravens as they shot past him.

Gorillas in the wild are inconsistent in their reaction to rain. We have noted elsewhere how gorillas in captivity love to frolic in the rain. Sometimes a group will squeeze up around the trunk of a tree to keep away from the raindrops coming through the leaves. On another day the whole group will simply sit out in the rain, apparently ignoring it. During the rainy season the gorillas seemed prone to catching colds. They often do a lot of coughing, and Schaller thinks that respiratory infections and related diseases are probably the main causes of mortality in wild gorillas.

Gorillas, as we have pointed out earlier, almost always walk on all fours and rarely walk on their legs for any distance. In most cases this does not exceed five or six feet. Schaller once saw a gorilla walk more than twenty feet bipedally. It was raining at the time, and he felt that the animal seemed to be anxious not to bring its hands and chest in contact with the wet vegetation. At another time when it was raining he saw a silver-back male walk bipedally about twenty-five feet with its arms folded. He saw a female who walked bipedally a distance of sixty feet until she reached a shelter under a tree.

Schaller did not supply the answer to how gorillas respond to a sick member of their group or to one who has died. In *The Ape People* a story was told by Prince Rainier of Monaco about two chimpanzees in his zoo. When the first animal died, its cage mate covered it with all the available straw that was in the cage. In other words, it was a rudimentary beginning to the ritual of burial of the dead. Schaller was not able to make any contribution to this subject from his own observation of wild gorillas, but the famous English anatomist Richard Owen has claimed that apes cover their corpses with heaps of leaves and loose stones which they collect and scrape up for this purpose. Schaller is not sure whether gorillas mourn the death of a group member or whether they simply abandon a corpse or the sick animal, leaving it to be devoured ultimately by scavengers and maggots. According to Gil Clayton, who worked in the Virunga range in Rwanda with Dr. George Frazier, the sick animal does tend to wander away from the group and may die in its absence. Perhaps this is why they do not actually see death in most cases unless the animal sustains death by accident, which must be rare in a gorilla group. It is possible that the custom of burying bodies has developed during the transition between apelike ancestor and man with the idea of preventing carnivores and predators from eating the animal after it has died.

Fred Merfield, who spent many years in Africa hunting gorillas, said, "Gorillas never abandon their wounded until they are forced to do so, and I have often seen the old man trying to get a disabled member of his family away to safety."

On one occasion an old silver-back male from a gorilla group had been seen in the vicinity of Kisoro, obviously very sick, but he had an infant with him. The silver-back male died, and the infant was subsequently captured by humans. It was a matter of conjecture why the animal had left the group. The fact that an infant was with it was a little puzzling, but the animal may have left the group with the big animal inadvertently, not realizing that the animal would subsequently die and that it would never again see its mother and the rest of the group. It was a very young animal, not yet fit to go off on its own into the jungle and rejoin its group.

Schaller noted a 23 percent mortality among babies born to a group on which he was keeping records. He subsequently estimated that about 40 percent to 50 percent of gorilla youngsters die before they reach six. In the wild they apparently have a very high birth rate. There were twenty-seven females in the group that he kept records on over a period of ten months. Only two among these twenty-seven did not have an infant or failed to give birth to one during that time. These two were obviously elderly and probably not fertile any longer. In the wild the gorilla is a very relaxed animal; placid is probably the best word for it. Its whole behavior is one of self-confident arrogance that indicates it feels itself beyond being threatened. In fact, it really has nothing to worry about except an occasional leopard, which usually will attack only an isolated and sick animal, and man when he comes with his nets and guns.

It has been mentioned that mountain gorillas never eat anything of an animal nature. They are quite different from chimpanzees, which will even kill and eat baboons, and a number of observers have seen chimpanzees in the wild catch young antelopes and make a meal of them. Jane Goodall has seen chimpanzees in Tanganyika eat monkeys and even indulge in cannibalism. Schaller found a record of a chimpanzee attacking a human baby:

> When I looked through the government files for the town of Kigoma, Tanganyika, I came on the following record of meat eating by chimpanzees. In March 1957, about 5 miles north of the town, a woman walked along a path carrying a child on her back. The inquest report describes in the woman's words what happened next: "Then suddenly from the bush came a chimpanzee. We were in the bush and the village was far. I was tying up my kuni [firewood]. I ran away and the chimpanzee bit me twice. He was about 4 feet tall; I fell down. Then it caught the

child who was on my back. I made a good deal of noise and other women came. We saw the chimpanzee eating the child's ears, feet, hands and head." The medical officer in Kigoma examined the body and found that the "scalp was missing and there were five depressed fractures of the skull. Both hands were missing, one half of the right foot was missing. The injuries were consonant with the child's being eaten by some animal. The fractures of the skull were caused by teeth."

Schaller, in one of his books, makes a number of comments about the importance of learning more about animals and particularly of studying primates in the wild. One of his comments is as follows:

> There is another reason for studying animals, to me the most important of all: that man learns to understand himself. In an age when man is increasing his mastery over nature and is occupied with devising new weapons for which to obliterate himself, he is as yet only on the threshold of learning the causes underlying many of his actions. Because the behavior of man is so modified by the culture into which he is born, it is often easier to obtain a clearer perspective of some problems of human behavior by studying other animals. The closest relative to man and a group to which he belongs, are the primates—the lemurs and other prosimians, the monkeys, and the apes. Clues to the origin of human behavior and human society might thus be found in various animals and especially in primates.

The work which Dr. Schaller carried out in Africa is a classic study of gorillas in the wild. Most of it was carried out in the eastern Congo, but in West Africa Jorge Sabater-Pi had been making studies of the behavior and particularly of the dietary habits of the lowland gorillas in Río Muni, in equatorial Guinea. His publications provided a good framework for comparing the western gorilla with the eastern mountain gorilla. Unfortunately, in 1969 there was a great deal of anti-Spanish sentiment in that part of West Africa, and Sabater-Pi had to leave and return to Spain without having the opportunity to pack all his books and papers and records. Sabater-Pi was subsequently assisted by Clyde Jones, now at the San Diego Zoo, and they made a number of interesting observations on the lowland gorilla. One of the things they found was that since the gorillas were so wary, it was necessary to wear drab clothes when they were working in the field. Like Schaller's, their body movements had to be deliberate and slow, and it was necessary to make as little contact with the vegetation as possible because of the noise. It was also desirable to communicate by hand signals rather than by calling out and making a noise which was certain to disturb the gorillas. If vegetation had to be cut, it was important that it be held firmly and cut slowly.

In Río Muni there are fewer than eight people per square kilometer, and there is very little forest that has not been disturbed; a lot of it is secondary jungle. Most of the primary forest is confined to areas where the terrain is difficult, such as in mountainous areas, and to the margins of rivers, which are not suitable for farming and are not accessible for timber exploitation.

The western gorillas in Río Muni are mostly concentrated in a limited number of small geographic areas; some of these are limited by rivers and by the activities of humans (especially lumbering or agriculture) in the neighborhood. In Rio Muni there are about .58 to .68 lowland gorillas per square kilometer of area. The groups which range about the area vary between two and twelve animals.

Sabater-Pi made his first observations of the lowland gorillas in the wild toward the end of the 1950's, and he gave details of the stomach contents of a number of lowland gorillas which he had obtained in Spanish Guinea. His study showed that at certain times of the year the diet of gorillas is composed almost entirely of banana pith. They obtain this, of course, by stealing the banana plants from plantations belonging to the Africans. On other occasions the stomach contained nothing but the remains of forest fruits. The wild gorillas of Río Muni seem to live completely on vegetable material. A plant called aframomum represents about 80 percent to 90 percent of the diet of gorillas in this part of Africa throughout most of the year. They also eat great quantities of buds from trees, and the pith and bark of quite a number of plants are also popular with them. Sabater-Pi made a list of the different kinds of plants that the western gorillas would eat. It totaled up to seventy-nine varieties.

Río Muni gorillas and chimpanzees use the same area of the forest, but rarely seem to come in contact with each other. Gorillas, however, tended to live on the edges of the forest, preferring the secondary or regenerating forests, whereas the chimpanzees were more inclined to dwell in the deeper parts of the forest. So while eastern gorillas were almost completely terrestrial, only the young gorillas climbing trees, the chimpanzees were arboreal as well as terrestrial, and this also helped to make

198

for a separation because the chimpanzees tended to occupy areas where the trees were bigger and stronger. The western gorillas as seen in Río Muni were very timid animals and did not show any aggression, although they were prepared to cross a road if it was on their trail. They would also cross small open areas, but generally they liked to be secluded in the areas of secondary dense vegetation. Sometimes Sabater-Pi and his colleagues got as close as thirty feet or even less to the gorillas, and although they were aware of the presence of humans, they never moved directly toward an observer. Twice contact was made with gorillas at a distance of only three to ten feet (one to three meters approximately). These were adult males, and in each case the only reaction of the animal was to escape. Nevertheless, the local people in Río Muni have many stories about the fierce attacks which gorillas are supposed to make on humans.

There was a good deal of gorilla hunting in this area. Sabater-Pi sad that between May 28 and June 1, 1968, four gorillas had been shot by hunters in the Mount Allen region. Another animal had been captured in a snare trap, but because it had no commercial value and because it had a severe injury caused by the trap, it was butchered. The Africans in this area believed that hunting gorillas was desirable and necessary to keep them under control because they raid their crops. A good deal of the destruction of apes in Río Muni, however, was apparently due to the collection of gorillas for sale, both legally and illegally. The method common in other parts of Africa was used to acquire baby apes; that is, the hunters killed the adult female in order to get the infant gorillas or chimpanzees. In the ten years before 1968, Sabater-Pi said, at least forty gorillas and sixty-four chimpanzees were exported legally from Río Muni, and many more of the habitats of the gorillas were being destroyed. In scattered parts of the country the forests were being cut down so that the lands would be available for farming. Approximately one-third of the land was being exploited for the lumber industry.

In the early morning of October 1, 1966, in the jungle of Río Muni, there was a light fog drifting through the trees. Suddenly from a small clearing came two shots from a gun. The villagers rushed from their huts to the banana grove, where Benito Mané was standing with a gun. Lying at his feet was a dead black gorilla; in its hands was a banana stem, torn apart, so that the gorilla could get at the pith, which it especially liked to eat.

199

Mané belonged to the Fang tribe, and he had worked long and arduously to develop his plantation. As the villagers arrived, Mané indicated to them a small white-haired creature clinging to the dead gorilla with its head buried in the black hair, and he called out, *"Nfumu ngi!"*—the translation of which is "white gorilla." It was the first white gorilla known to science. Rumors of the existence of white gorillas have occurred from time to time, however, and in his book published in 1896 Professor Robert Garner mentioned a report about white gorillas:

> The white trader living on this lake claimed to have seen a gorilla which was perfectly white. It was seen on the plain near the lake. It was in the company of three or four others. It was thought to be an albino, but in my opinion it was only a very aged specimen turned gray. A few of them have been secured that were almost white. It is not, however, such a shade of white as would be found in an animal whose normal color is white. I cannot vouch for the color of this ape seen on the plain, but there must have been something peculiar in it to attract so much attention from the natives.

Other gorillas have been found in this area which lacked pigment in the hands and feet, but these slowly colored up and became normal by the time the animals were twelve to eighteen months old.

Mane picked up the little white gorilla and took it to his hut and put it in a cage, which he lined with vegetation, including all the leaves and ferns and sticks that a gorilla would use for building itself a nest. The little gorilla with white hair and white face was not a complete albino because it had blue eyes instead of the pink eyes of a true albino. The baby was fed stems of plants and various buds and wild fruits. He was kept in Mane's hut four days; then he was taken into the nearby town of Bata, where he was purchased by Sabater-Pi. At that time Sabater-Pi was the head of the animal acclimatization station of the Barcelona Zoo, and he set to work to tame the little gorilla.

When the baby arrived in Bata, he was covered with red dust from the road, so he was promptly given his first bath. He weighed at that time nineteen and one-quarter pounds and was about two years old. When he arrived at the animal acclimatization station, he was given a new cage and a pan of milk, which he drank with gusto, and so his acclimatization began.

Because of the lack of pigment in the eyes, the baby had what is called photophobia and would close his eyes if exposed to bright light. Even in diffuse light, similar to that in the forest where he lived, he still blinked about twenty times a minute. He was very slow at responding to human contact at first, but by the sixteenth day he had allowed his head, arms, and legs and his back to be touched and was allowed to leave his cage. At that time he followed anyone who showed him a food he liked, and these included, by this time, bananas, sugarcane, cookies, and milk. Every morning and afternoon he was given personal contact with humans for an hour in order to facilitate the taming process. At the end of a month Snowflake would walk around hand in hand with any of the people whom he had come to know. He would play very happily, clapping his hands and turning somersaults. He followed Sabater-Pi and his wife everywhere. At that stage he was ready to go to Spain and be exhibited at the Barcelona Zoo.

In due course the little white gorilla was sent off to Spain, but before leaving, he was given tonics and treated for parasites. At the zoo Dr. Antonio Jonch Cuspinera called him Capito de Nieve, which means "snowflake" in English. Snowflake was the subject of intensive scientific study by Dr. Arthur Riopelle, a former director of the Delta Regional Primate Research Center in Covington, Louisiana. Dr. Riopelle had arranged that Snowflake would be brought up with a normally pigmented gorilla called Muni, a black male of the same size and age. Snowflake very much enjoyed the crowd; in fact, he was described as a ham and behaved rather like a circus clown. He seemed to enjoy people as much as they enjoyed him. Of the two gorillas, although Snowflake was the one that played to the human audience, Muni was the one that took the lead in the gorillas' own games, leading their various explorations. At six years of age Snowflake weighed about 100 pounds and had become very affectionate, but he was still liable to throw temper tantrums. When he was frustrated, for instance, by an intelligence test that he was not able to solve, he screamed, slapped the sides of the cage, and tried to destroy the testing apparatus. Dr. Riopelle recorded the physical and behavioral development of Snowflake so that he could be compared with other gorillas and felt that Snowflake was getting a bit spoiled. An extensive photographic record was made of the behavior of Snowflake and his cage mate, Muni. At one time they were given a human doll, and when the doll was put into the cage, at first they simply sniffed at it in a tentative fashion, touched it nervously, and then finally picked it up and swung it around in such an excited fashion that eventually they tore off its limbs. Snowflake was quite agile and light on his feet. On one occasion he was sitting on the top of a jungle gym when it started raining, and he stuck out his tongue just like a child to catch the drops. When Snowflake and Muni peered through the glass window of their cage at the spectators looking at them, they often shaded their eyes with their hands.

Gorillas are not naturally carnivorous, as we have pointed out, but both of the cage mates would eat ham and chopped beef as well as fruit. They also enjoyed a jar of yogurt occasionally. Both Snowflake and Muni have now developed large appetites. Their main meal, which was served at about 1:30 P.M., provided for each animal two pounds of bananas, one pound of apples or pears, one-half pound of quince jelly, one-half pound of boiled ham or roast chicken, and one-quarter pound of bread. There were three lighter meals, and these consisted of raw beef, yogurt, hard-boiled eggs, rice, and cookies. So they are really living in the lap of luxury.

It is possible that the sensitiveness of Snowflake's lightly pigmented eyes caused him to lose some confidence, and this may be the reason why Muni dominated him in most of their joint activities. Muni also exhibited a great deal more confidence when a mirror was introduced to the cage. He was not frightened by it and paid great attention to his reflected image. On the other hand, when the mirror was first introduced, Snowflake ran from it, and on the second introduction he showed both curiosity and fear. He bared his teeth at it and beat on his mirrored reflection. When he was introduced to a six-year-old boy wearing a gorilla mask, Snowflake studied him for five minutes and then decided that the mask was a fake and ripped it away from the boy. He also was able to solve quite complicated psychological problems. When a young female, normally colored gorilla was put in the cage, Snowflake asserted his authority and repeatedly pushed Muni away from the little female. Because of this interest in the female, it seemed obvious at that time that he was developing along normal lines sexually. In fact, he has now mated successfully with a black female gorilla and they have produced a black baby; that baby, however, is carrying the recessive genes for whiteness, so in due course, by mating two of Snowflake's children, the Barcelona Zoo will probably produce white gorillas. The New York *Times*, when Snowflake was seven years old, described him as the "vanilla gorilla." The paper provided a number of interesting pictures of him playing with Muni and playing and wrestling with a female black gorilla called Afanengui.

The general result of the psychological testing that was done on Snowflake showed that he was no brighter than other apes, although he was certainly not any duller.

Sabater-Pi estimated that there were about 5,000 gorillas in Río Muni in 1964, two years before Snowflake was found, and that they were dispersed in four major local population groups with a total of about 1.3 animals to a square kilometer. It should be mentioned that the studies on gorillas in Río Muni were financed by the National Geographic Society, which has done much for the advancement of the conservation of nature throughout the world.

9

The Real Gorilla Stands Up—He Is a Peaceful Vegetarian

GEORGE Schaller's work in the field over a period of fifteen months has been called basic to our knowledge of gorillas in the wild, but the final identification of the true gorilla nature was the work of an intrepid young investigator named Dian Fossey. In 1964 she was a physiotherapist working at Stanford University, and her dream was to go to the Congo and study wild gorillas. Her wish was finally realized, and in January, 1967, she set off for the Congo via the subdepartment of Animal Behavior at Cambridge University in England. She received major financial support from the National Geographic Society, additional financial help from the Wilkie Brothers Foundation, and scientific help from Dr. Louis Leakey.

Dian Fossey began her work in the Congo on the slopes of Mount Mikeno, like many distinguished explorers before her, but was able to stay there for only six months because of political problems in Kivu Province. This was unfortunate because the gorillas in Kivu were in a well-protected park system and were not constantly threatened with human intrusion. Consequently they were not frightened by Dian Fossey's presence. After leaving the Congo, Dian relocated in Rwanda, setting up her new camp in a large meadow that was part of the saddle between mounts Karisimbi and Mikeno and Visoke. The gorillas in that area, however, had been so harassed, by both poachers and cattle grazers, that she had great difficulty at first in making contact with them.

In our development of knowledge about the gorilla, we have built up a picture, described by many hunters and explorers, of the wild, ferocious, vicious, unpredictable gorillas—particularly the mountain gorilla. Then came George Schaller, who documented the fact that these were not aggressive or ferocious beasts and that one could walk among these animals without being attacked, especially if one left his gun behind and approached the animals with appropriate humility and no arrogance in his walk or behavior. And, finally, much to the surprise of the still dubious, Dian Fossey showed what a young, unprotected woman can do among these alleged "ferocious" and "vicious" animals.

Dian Fossey worked with the gorillas for three years in her first field investigations and then for another eighteen months. During this time she got to know many of these animals as individuals, and several of the groups would accept her presence almost as if she herself were a member. She could approach within a few feet of many of them, and some of the animals, especially the juveniles and even young adults, would come close enough to her to pick up her camera strap and examine it, to play with the buckle of her knapsack, and even to manipulate the laces on her boots. She got to know some of the animals so well that she gave them names and was able to recognize many of them after not seeing them for periods of time. The former methods of sitting and observing gorillas which most investigators had used were not enough for Dian. Gorillas, she found, become suspicious of anyone alien to them who simply sits around and stares. So Dian found it an advantage to pick up their habits and gestures and incorporate them into her own behavior. Dian imitated gorilla feeding and grooming behavior and later, when she got on more familiar terms, copied their vocalizations. She even managed to duplicate what she described as startling, deep belching noises. The gorillas seemed to accept this type of behavior—even though it was not very "ladylike." One of the important gestures both Schaller and Fossey learned to use when contacting gorillas is folding the arms. Apparently the gesture suggests to a gorilla an act of submission or at least an indication that no harm is meant. Dian claimed she often felt rather stupid thumping her chest rhythmically like a gorilla or sitting about in a patch of wild celery pretending to munch on it. But it enabled her to be on the closest of terms with the giant apes.

She established an observation camp, a sheet-metal cabin, at 10,000 feet on Mount Visoke and made her first advance into it when there was fog in the area and it was wet and very uncomfortable. It was not long before she acquired two young gorillas. They had been captured by Rwandese park rangers and tribesmen for a zoo in a European city, despite the fact that the International Conservation Organization had designated the mountain gorilla as a rare species and stated that the animal should not be captured. The young gorillas were in such a bad state that Dian Fossey, though she was concerned about the reasons for their capture, made an offer to take care of them until they were well enough to be shipped to the zoo in question. One of the animals, Coco, had been sitting for twenty-six days in a wire cage that had not allowed him room either to stand or even to sit up. He had received no liquids and had been offered foods to which he was not accustomed and which he refused, but he had accepted bananas and this is what had saved him. Bananas alone are far from an adequate food, and he was very thin and weak and, as a result of his experience, was intensely scared of humans. Puckerpuss, the other animal, had refused to eat anything. She was also very thin and weak and had a great fear of humans. Dian spent a few weeks getting to know the young animals and giving them medication and new foods. In the midst of this she acquired another problem. Her cook quit when she asked if he would help her prepare the bottle formula incorporating dried milk and other preparations that she was using to feed the gorillas. He told her in Swahili that he was a cook for Europeans and not for animals. So she had to give up her observations for the time being in order to look after the two young babies. When they got better, she took them out to the forest to watch their feeding habits and their grooming and to listen to their vocalizations at close quarters. She was intrigued by the maneuverings of the animals when they were searching for worms and beetles under the bark of tree trunks and how they themselves were looking for small pieces of scurf. Then finally they were well enough to be sent to the zoo and she felt very sad at losing them. This is the first evidence we have had that gorillas eat anything other than vegetable food.

At this time Dian was making an intensive study of four groups of gorillas. When she came across one of the groups that she knew and that knew her, she used a particular type of vocalization that she had learned from the two little gorillas she had picked up. This vocalization apparently meant, "Food is served. Come and get it." She said the reaction of the group was immediate, and the leader came up to her with an expression on his face which suggested, "Come on, now. You can't fool me." In this particular group there were no females or infants. There were five males, and since they did not have to worry about protecting anybody, they did not repress their curiosity about anything. Sometimes she climbed a tree to observe the group, and then three of the gorillas would climb up the tree and join her and spend a good deal of time investigating her camera equipment and her boots and clothing. When she first met this group some time ago (two years before), they were not an exclusive bachelor group at all, but they had an elderly, tottering female living with them. This animal's arms were very thin and apparently the muscles were atrophied. Her breasts were dried up, and even her head was graying, and Dian Fossey estimated her to be about fifty years old. The five males seemed to be very attracted to her, and most of the activities that were carried out by the group seemed to center around this aged animal. She would always be the one that would introduce mutual grooming, and they would quickly form a sort of daisy chain of gorillas—each intently grooming the other.

Eventually this elderly female died, but before she finally disappeared, she gave signs which could be construed as being senile by wandering away and doing a kind of aimless circling in the forest. When this happened, the five males were very patient and would just wait for her to come back again, although sometimes a soft bark from one of them would stimulate her to rejoin the group. When she arrived, she would go to the leader of the group and embrace him—a gesture which he would usually return. These two would sometimes share the same nest at night. One day the old lady and the leader of the group went off into the jungle and were absent for two days. At the end of that time the leader came back, but he was alone. When Dian backtracked along their trail, she found that the two of them had shared nests together, but she was never able to find any evidence of what happened to the old female or to her body.

The behavior of the group was often related to the character of the leader, and in one of the groups there was a silver-back leader that she named Whinny. This was a very calm animal with an imperfect ability to vocalize. He got sick and was ill for some months and finally died. Then the leadership of the group was taken over by an animal, also a silver-back, that Dian Fossey called Uncle Bert. He reversed the gorillas' rather placid acceptance of her presence, and instead of the animals relaxing and accepting her when they saw her, they would engage in an orgy of chest beating and hitting the foliage. The new leader also changed the trails that they took. Although she thought of Uncle Bert as a "catankerous old goat," on one occasion toward the end of a rest period she saw a small infant gorilla approach Uncle Bert and lean against him. She had thought Uncle Bert's reaction would be one of annoyance, but instead he picked up a flower and actually tickled the baby gorilla with it. Very soon the baby scampered around like a puppy, but Uncle Bert was still lying on the slope holding the flower in his hand with "a most idiotic grin on his face." Play appears to be a very important part of life in a gorilla group, particularly for the young animals. One sequence was recorded in a group where a play period lasted more than twenty minutes with a great deal of rolling, sliding, and tumbling down the slope, grabbing branches and swinging on them until they broke, and generally tangling with each other in an orgy of fun.

A gorilla group will take extreme measures to protect the young. The three animals in one group that were being observed sometimes had paternal inclinations, and they would pluck an infant from the mother's arms to groom it. The older infants and also the juveniles seem to feel very secure in the protection of the adults and therefore will often carry out actions which will irritate the big silver-back males until the limits of their patience are reached. One day a little gorilla Dian called Icarus was trying out a new acrobatic routine. Icarus climbed up a small tree and was swinging by his arms when the branch suddenly broke and came crashing down. The group seemed to blame Dian for the accident. The whole group charged her, led by the males, with the females following, and they stopped only five to ten feet away, apparently when they saw that the juvenile was perfectly all right and climbing another tree. The males, however, remained very tense and continued to give barks of alarm. Then she was dismayed to see the small infant gorilla climb the same small tree which had broken and begin a series of acrobatic adventures on it. The silver-back males looked at the baby and looked back at her; whenever their glances met, the silver-back males roared their disapproval. Finally Icarus climbed up into the tree and started a game with the baby, which eventually led him back to the group. With a sigh of relief, Dian found that the crisis was over.

Dian also stresses the gorilla's lack of aggressiveness and says that in 2,000 hours of observation she saw only about four minutes of behavior which could be called aggressive. Most of this was bluffing behavior. The closest call she had was when five large roaring males charged at her. The leader was only three feet away when she spread her arms wide and shouted,

"Whoa." And the whole group stopped. She described mountain gorillas as nothing but introverted, peaceful vegetarians, a judgment which is undoubtedly correct.

Dian Fossey refers to the many reports of gorillas attacking humans when the latter hunted them. She talks also of the "intrepid white hunters" who have been very courageous in facing the charges of these screaming black monsters and says that the picture given by them is quite false. Dian Fossey was even able to make physical contact with one of the animals in the group which she was studying. An animal she called Peanuts once approached within a few feet of her while she was sitting and began to show off in front of her by beating his chest and strutting around. Then he moved to within a couple of feet of her. She gestured in a way she believed the gorillas find reassuring and began to lift one arm and scratch herself under the other arm. She was delighted to see that Peanuts imitated her. She then offered her hand, palm up, which she thought the animal would recognize as being like his own palm. Peanuts stood up, was uncertain for a moment, and then finally touched her hand with his fingers. This is probably the first recorded friendly physical contact between human and wild gorilla. Such an act certainly makes the brave hunters of the past look foolish. Peanuts was excited at this gesture because he walked away from Dian, beat his chest, and went racing off to join his group, which was feeding in a calm fashion about eighty feet away up the hill. Dian was so excited by this incident that she could not help crying. Since she was close to leaving the area to go to Cambridge to work up her material for her doctoral thesis, she felt that this was "the most wonderful going-away present I could have had."

In an interesting vignette she describes how gorillas can be both gentle and have a sense of mischief. She said,

> One day Bravado, a young male, tried to climb past me down a tree trunk where I had settled myself on a limb to observe and take some pictures. Bravado made his way up easily enough, brushing past me as if I were not there. But on the way down, he apparently decided I was in the way and should move. Once his head filled my view finder, I decided it was time to turn my back to him. Just as I got a good hold on the tree, I felt two hands on my shoulders, pushing down.
>
> I had often seen gorillas do this to one another when they wanted the right of way on a narrow trunk. Not wanting to risk a fall, I refused to budge. After another moment of gentle

pressure—only a fraction of the mighty shove he could have given me—Bravado moved back. He beat his chest, then jumped out onto a side limb. He hung there by two arms, bouncing deliberately, knowing that his weight would break the branch and thus provide a satisfactorily loud snapping noise. He succeeded; the branch broke with a crash and Bravado landed eight feet below, where he calmly began feeding.

Each of the nine groups that Dian had eventually observed was ruled by a dominant male silver back who apparently had unquestioned authority. There were sometimes one or more subordinate silver backs according to the size of the group; there were also younger black-back males, which were usually mature and had not yet developed the silver color on the back. Other members of a group could be females, juveniles, and infants, although all-male groups could occur.

Mountain gorillas have three vocalizations that are very clear. One is called the belch vocalization, which Dian Fossey had learned herself, and she found that when she used it, it usually brought responses from any animals nearby. Once she managed to crawl into a feeding group of gorillas and start the belch vocalization, and the animals around her answered it. This vocalization apparently expresses a feeling of comfort and well-being; the same vocalization with the same message figures in the vocal repertoire of human primates.

Another type of vocalization is called the pig grunt. The pig grunt is a harsh staccato sound and it is used when discipline is required—for instance, when a silver-back male has to stop a group of squabbling females or if he wants to move on and warn the group to get up and follow him. Females use a softer version in the discipline of their babies. Then there is the whoop bark, which is actually an alarm vocalization, and it is used when the animal is curious. If a silver back gives this alarm bark, then the whole group is attentive to what is going on. Dian spent many hours recording the various vocalizations of the gorillas and is now carrying out further studies of the various tapes that she has taken of them.

We mentioned earlier that in one group the leader, Whinny, died and that Uncle Bert, who seemed to be a rather catankerous animal, took over the responsibility. Uncle Bert was very strict and domineering, particularly at the beginning. One of his first acts was to eject a subordinate silver back from the group altogether. Dian had called this silver back Amok. For a time she thought he might join another group, but eventually he became a lone animal, leading a solitary life and simply wandering around the jungle in what appeared to be a kind of haphazard fashion. She was not able to make much contact with him; whenever she saw him, he would give an alarm vocalization and then run away. On one occasion the gorilla Peanuts, with whom Dian had had previous contact, popped up behind a log just near her. He looked at her with what she thought was an impish expression. She had in her pocket three cherrylike fruits from a tree that gorillas liked, and she put one on top of the log, expecting the gorilla to back away. On the contrary, he came forward, grabbed the fruit, and put it in his mouth. Then he looked her in the face as if he were asking for more fruits. She took the remaining two fruits from her pocket and extended them to him on the palm of her hand. He extended his hand and grasped them without any hesitation and popped them, too, in his mouth. He then waited to see if she could produce any more, but she had pocketed only those three, and Peanuts moved around to her other side, leaning toward her, and then whirled around and ran off back to his group. This was the second close contact she had had with him. This was really the final proof. There was no doubt the gorilla was simply a peaceful, introverted, gentle vegetarian, a gentle giant, in fact.

In October, 1971, after a term at Cambridge, Dian returned to her cabin on Mount Visoke with four young recruits and some assistants from Rwanda. She was planning a census of the gorilla population, partly to find out how many there were and partly to underscore the necessity for protection of the mountain gorillas. When she returned to camp, she was horrified by the news that there had been a slaughter of gorillas just to the south of the area where she had been studying the animals. The bodies of five gorillas had been found scattered over a distance of seventy-five yards. They had been mauled by dogs; they had obviously been injured by spears and had been hit with stones. As far as she could tell, this had been done simply for the excitement of hunting gorillas—a horrible sport engaged in only by the highest of primates, man.

Dian Fossey's initial census of the mountain gorilla has been followed up by Dr. A. F. G. Groom of the University of Cambridge. The census has also been supported by the National Geographic Society and by the World Wildlife Fund. The main area of interest in the census of gorillas is around the Virunga volcanoes. It appears that if there are continued invasions of the sanctuary by humans at the rate which was found in 1971, eventually the six mountains in the chain of volcanoes will become isolated from one another because of the activities of the humans on the lower slopes of the mountains; this will probably confine the gorillas solely to the peaks, and if this happens, they will not survive. In 1972 there was another census in the six mountains in the park, and one of them was found to have no gorillas at all and another had only a minimal population which showed signs of being harassed. In one of these areas when a gorilla was seen, it ran screaming into the jungle as if it had come to regard man as a very serious enemy.

In 1973 Groom led a group of four people into the Virunga volcanoes, and on the easternmost mountain, Muhavura, they found only thirteen animals, even though the circumference of its base is much greater than that of Visoke, where there were possibly as many as ninety-six animals. They found no gorillas on Mount Mahinga. On Sabinio they found thirty-two and on Mount Karisimbi eighteen gorillas, divided into three groups. When they saw observers, a number of gorillas either screamed and ran away or just silently took off with all the evidence of having developed fear of humans. Only on one mountain, Karisimbi, did there seem to be no signs of fear. The animals would give a bark of alarm and then retreat for a little distance. One of the factors which resulted in Mahinga and Muhavura having few or no gorillas was the presence of smugglers' routes which ran between Uganda and Rwanda. Apparently the volume of traffic is considerable. On the Sabinio-Mahinga route there was some evidence that groups of ten to twenty smugglers passed along about every half hour all through the day and that the second route, which actually ran between Mahinga and Muhavura, was used only a little less. The smugglers make a lot of noise when they travel. Apparently they do this to scare away larger animals, which presumably might attack them, but they certainly disturbed the gorillas also. These areas are also invaded by woodcutters, who remove the trees, and grass cutters, who use the grass for thatching houses. The penetration of herdsmen with their cattle into these areas is also very damaging to the gorillas. In Uganda the designation of a reserve for gorillas simply meant to the local population that they could do what they like in the area. At the time of the 1973 census Groom and his colleagues found no more than twelve gorillas in the whole Ugandan part of the gorilla sanctuary. They came to a number of conclusions, and among them are the following:

212

If the mountain gorilla is to survive in the area, territorial encroachment must cease and the human interference be reduced. In other words, the status of the Virungas as a national park or reserve must be respected. Zaire signed (but has not yet ratified) the 1968 African Convention for Conservation of Natural Resources, under which the gorilla is totally protected, and the signatories concerned have undertaken to protect its habitat. The problems of enforcing national park or reserve status are aggravated, because the African officials do not have sufficient motivation to carry out their duties. Poorly paid, they are much more inclined to accept bribe money from the poachers and cattle raisers who can afford substantial sums. Poachers rarely seem to attack gorillas directly, but the witchcraft motive is still strong. The dead silver-back which we found attacked on Muhavura and moved to Miss Fossey's camp at Visoke had its ears, tongue, genitals and the terminal phalanx of each little finger cut off on the journey, even though we had paid guards to prevent interference. Our trackers assured us that these parts of a large gorilla were powerful witchcraft weapons against an enemy. In the 1971 census, we had found a silverback which had apparently been killed expressly for such purposes.

So it is apparent that a great deal must be done to protect the mountain gorillas and to enforce this protection if they are to be saved for posterity.

In the Republic of Zaire, there is a determination to protect the mountain gorillas of Bukavu. President Mobotu Sese Seko has said, "We in Zaire are never disappointed not to be able to show our guests old cathedrals or monuments because the heritage of our ancestors is the natural beauty of our country, our rivers, large streams, forests, insects, animals, lakes, volcanoes, mountains and plains. In a word, nature is the integral and real part of our originality and personality."

Near Bukavu at the south end of Lake Kivu the Zaire government has established in the east of the country a wilderness sanctuary 600 square kilometers in area and called the Kahuzi-Biega National Park. This is the home of 250 mountain gorillas. The warden of this park is Adrian Deschryver, who has been given absolute authority over it. He tries to keep the nearby African tribesmen from using the park to graze their cattle, a depredation which would soon encroach seriously on the reserve and endanger the gorillas. He warned one tribe several times about this, but when they persisted, he finally had the cattle shot. Deschryver also keeps an eagle eye open for poachers, and he is confident that this evil has been practically eliminated.

John Heminway paid a visit to the Kahuzi-Biega Park to decide if it was possible to make a film for television on the gorillas of Bukavu. There are two groups of gorillas in the area, which Deshryver has habituated to the process of being continually observed. He now takes small groups of visitors into the jungle to observe them.

On his first sortie into the jungle with Deschryver, Heminway was faced with a charge by a male gorilla. He commented that he smelled the silver-back male long before he made the charge, which was preceded by a bark. The bark soon became a shriek, and then, in a flash, the gorilla had run past them and stopped fifteen feet away. What struck Heminway after this was the unearthly silence. The gorilla had disappeared; there was not a sound, but he knew that they were both being examined minutely through the trees by a 400-pound male gorilla that they could not see. Heminway saw other gorillas before he left, convinced he must make a film about Deschryver and his work, and he says, "Our film will not mock gorillas. We will not humanize them; we will give them no endearing names, nor create a fiction of their lives. We will not be tempted to make them appear more dangerous than they truly are. We will honour Zaire, its national parks, the men who work for them, and the freedom of these great animals which they protect." His film, which was shown on national U.S. television, did just that. It was a fine production and has, we are sure, helped to bring to the general public the importance of preserving for posterity these noble animals in their homes.

10
The Anatomy of a Gorilla

IT has been estimated that a gorilla has the strength of sixteen men, so you can imagine what would happen if a gorilla got into the ring with a heavyweight boxer. Would one of the greats of the past like Jack Dempsey land a punch or would he be out of the ring with one crushing blow? Unquestionably the latter would happen. Carl Akeley compared a mountain gorilla with Jack Dempsey, the former world heavyweight champion: "Although not as tall as Dempsey, the gorilla weighs nearly twice as much and his arms are longer and more powerful." If it came to running away, however, Dempsey would have been safe, for Akeley goes on to point out that "the gorilla's legs are much shorter and that unquestionably, a well-developed man can travel both faster and farther than a gorilla."

Here is how the gorilla and Dempsey compare:

	GORILLA	DEMPSEY
Height	5 feet 7⅓ inches	6 feet 1 inch
Weight	360 pounds	188 pounds
Chest	62 inches	42 inches
Upper arm	18 inches	16¼ inches
Reach	97 inches	74 inches
Calf circumference	15¾ inches	15¼ inches

Just how big is the gorilla in the wild? Since its discovery more than 100 years ago there have been reports of animals attaining enormous size. This is not surprising. Most of the people who hunted gorillas in the nineteenth century and the early part of this century were, of course, anxious to claim that the animals they shot were especially large as well as aggressive, for

215

this made the execution of the animal seem so much more impressive and justifiable. So in the record book we find a number of claims for very big gorillas; some hunters' imaginations claimed there were gorillas seven to nine feet or more in height. There is very little possibility that gorillas of that size exist.

A German trader, H. Paschen, in 1900 in the Cameroons shot a gorilla and had a picture taken of the animal sitting on the ground with Africans alongside it. It was skinned and stuffed, and the specimen was exhibited first at Hamburg and later was purchased by Lord Walter Rothschild and placed in the Rothschild Museum in Tring in Hertfordshire, England. The original description of the animal was that it was of gigantic size and that its length when it was lying down was 6 feet 9½ inches with an armspread of 9 feet 2¼ inches. It was found, however, that its length lying down was actually taken from the crown of the head to the end of the toes and not from the head to the heel, which it should have been. Furthermore, if its armspread was 9 feet 2¼ inches, its standing height could have been only a little over 6 feet. Dr. W. T. Hornaday of the American Museum of Natural History claimed that the animal shot by Paschen was only 5 feet 6 inches in height, and Hornaday estimated its weight to be 500 pounds.

An article in a publication called *The Living Animals of the World,* edited by C. J. Cornish, claimed that this particular gorilla was only 5 feet 5 inches in height and had an armspread of over 8 feet and weighed 400 pounds. In Brehm's *Tierleben* the height of the same gorilla was said to be 6 feet 7 inches and its weight 551 pounds. These various figures for the one animal indicate how dubious we must be regarding claims for very large gorillas. Adult males from both races of gorillas vary from about 5 to 6 feet in height, and there is no specimen preserved which is over 6 feet in height. In the French periodical *L'Illustration* of February 14, 1920, a gorilla photograph appeared and was the basis of a claim that this animal was 9 feet 4 inches in height. The picture certainly suggested that the animal was very large, and it could be compared in size with the African who was sitting by its side. Assuming that the man was 6 feet in height, the picture actually indicates that the sitting height of the gorilla would be about 43 inches. This would have suggested a standing height of only 5 feet 8½ inches, which is quite a decent-sized gorilla, but certainly not one 9 feet tall. The French periodical *La Nature* of July 29, 1905, illustrated a gorilla which it said weighed 770 pounds and was 7 feet 6½ inches in height. It was even said to be 3 feet 7 inches from one shoulder to the other. Again, comparing this gorilla with the Africans standing and sitting with it suggested that the animal was no larger than the average-size lowland gorilla. These kinds of wild claims are unscientific enough, yet some of them have been repeated in such respected works as R. Lydekker's *Wildlife of the World* and Brehm's *Tierleben*. Gorillas are not, in fact, able to stand in a fully erect position as a man does, and therefore what is described as the standing height in gorillas is simply the length measured from the top of the head to the heel when the animal is in a lying position. There is a gorilla group in the American Museum of Natural History in what is called the Akeley Memorial African Hall. One of them Akeley described as "the lone male of Karisimbi." Akeley took the height of this animal, which he measured after he had shot it, and it was 5 feet 7½ inches. Its reach from one fingertip to another of the outstretched arms was 97 inches.

Dr. Adolph Schultz, a Swiss anthropologist who lived many years in the United States and made many distinguished contributions to anthropology, has measured the skeletons of ninety-three adult specimens of lowland gorillas. He has come to the conclusion that the average standing height of these animals is only about 5 feet 3 inches. The smallest adult was 4 feet 10 inches and the largest 5 feet 9¼ inches. Those are the figures for males. The average for females was 4 feet 3¼ inches. The heaviest mountain gorilla known was one that was in the San Diego Zoo and was called Mbongo. He weighed 660 pounds at the apex of his weight. His height was 5 feet 7½ inches. The average standing height for the mountain gorilla seemed to be about 5 feet 6¼ inches in males and 4 feet 7 inches in females. In the wild the adult male, according to Dr. David P. Willoughby, averages around 458 pounds and the female 240 pounds.

Normally the color of the hair of the gorilla is black, and the skin also has the same color. The older male gorillas show white on the back. We have some young gorillas at Yerkes that are not much more than ten or eleven years old in which there is considerable development of white hairs on the back, giving them a gray appearance. In some cases the skin may take on a grayish or brownish tint, but in most gorillas it is jet black. Gorillas have certain characteristic sweat glands in the body which produce perspiration that is very strong and acrid. There is no mistaking it when you encounter it.

The gorilla head is very large; both brow and chin recede sharply, but in the brow there is a development of large ridges of bone over the eyes, which are called supraorbital ridges. Probably the most characteristic part of the head, particularly in the male, is the development of a big bony crest on the top, to which are attached the masseter muscles, which move the heavy lower jaw. On top of these muscles a thick connective tissue is stretched. This crest gives the male animal a frightening appearance and raises the height of the animal because it projects a few inches above what normally would be the top of the skull.

Gorillas have small ears set well back against the head. In this way they differ a lot from the chimpanzee and the orangutan and are more like the human. The eyes are mostly dark brown. The gorilla has very large nostrils covered by a thick, black skin. The nose itself is flattened, and the animal does not have nasal bones, as humans do.

The large canines are conspicuous, especially in the male, which has a large mouth. Gorillas also have two sacs under the skin in the throat region. They are relatively small in the gorilla, but in the orangutan are quite large. It is possible that in the wild they can be inflated to some extent and be used as resonators to enable the sound to penetrate for some distance through the heavy jungle. This is undoubtedly the reason for the existence of these sacs in the orangutan, but we have not seen any record of their being inflated in the gorilla, in the wild or elsewhere, and it may be that the inflation is so small that it is not obvious.

The gorilla's skull was first described in 1848 by Sir Richard Owen, a distinguished anatomist, who was at that time conservator of the Museum of the Royal College of Surgeons of England in London. Owen was impressed with the fact that the gorilla skull very much resembled that of the chimpanzee, and he concluded that the gorilla was really only a variety of chimpanzee. Since Owen's original classification of the gorilla on the basis of the skull, however, other anatomists have decided that the gorilla belongs to a separate genus.

The Museum of the Royal College of Surgeons has a number of links with the early history of the gorilla. The first gorilla skeleton to come to England came to the college in 1851, when Owen managed to obtain it from a Captain Harris. Later on there were a number of other valuable additions made to the gorilla collection in the museum. F. W. H. Migeod, who had traveled widely in Africa and obtained the skulls of two adult male gorillas, also donated them to the museum. The college also acquired through Lieutenant Colonel R. H. Elliot the skulls of mountain gorillas. The existence of these gorillas was first made known to science by a German engineer in 1902, who saw the animals in the vicinity of Lake Kivu and called them "the Kivu gorillas," so he antedated Captain Oscar von Beringe by one year. Two of the skull specimens that Elliot presented to the college were those of a very large male and a juvenile animal. Other mountain gorilla skulls collected also from the Kivu area found their way to the Natural History Museum in South Kensington in London. These came from another African traveler, Captain J. E. Philipps. Sir John Bland-Sutton, who was at one time president of the Royal College of Surgeons, managed to collect a series of skulls and skeletons of the lowland gorillas in the Cameroons, which is a beautiful demonstration of the growth and age changes which the gorilla skeleton passes through between youth and old age. Bland-Sutton eventually presented this group of skeletons to the college.

In 1926 the distinguished anthropologist Sir Arthur Keith described two breeds of gorillas. Sir Arthur was in charge of the field station belonging to the Royal College of Surgeons in the town of Downe in Kent, England. This farm is used as an experimental station for the college. Keith described gorillas as being of two races, the highland gorilla, or Kivu race, and the lowland gorilla, or Gabon race. He said that the Kivu gorilla lives up to 10,000 feet above sea level and has a much longer and denser coat of hair than the lowland gorilla. He makes an interesting point, however, that although there are marked differences in external appearances in these two gorillas, he was unable to find marks on their skulls which would reveal any structural difference between the two. There seems to be only one well-defined anatomical difference. There is a muscle under the chin in humans and apes called the genioglossus, which is attached to a little pit behind the region in the middle of the jaw where the two halves join. In the lowland gorilla this pit in the bone from which the muscle arises is large and fairly deep, whereas in the highland gorilla (mountain) it is usually shallow or not distinguishable. In man this pit is also absent. So at least from this trivial piece of anatomical similarity, man is closer to the mountain gorilla than to the lowland gorilla.

There are, of course, a number of anatomical similarities between man and gorillas or, for that matter, between man and the other anthropoid apes, but in the case of man, evolution was toward the further development of the brain; in the case of the gorilla, it was toward the further development of brawn. Gorilla fetuses look more like human fetuses than do those of the chimpanzee or the orangutan.

The gorilla foot is interesting. What is called the tarsal element (the bones that extend from the ankle to the toes) is very long. The four smaller toes are very short. In this respect the gorilla's foot is intermediate in anatomical position between the chimpanzee and the human. The great toe of the gorilla, although it can be lined up close to the other toes, can also be opposed to it, in other words, extended out at an angle, and can be used for grasping just like a thumb. The gorilla uses its large toe in two different ways: When it is scrambling about in bamboo thickets or over branches, the toe is extended at an angle from the foot and is used in grasping as the animal progresses along. When it gets down onto the ground and the foot is being used as a supporting and propelling organ, the great toe then swings over and aligns itself parallel with the other toes. The animal gets its forward impetus from the power of that big toe. Keith believes that the human foot in the course of this last stage of evolution passed through a gorillalike stage.

There is another anatomical similarity between man and the gorilla, and this is in connection with a muscle called the peroneus tertius. This muscle in humans has split off from the long extensor muscle of the toes and is associated with the raising of the outer border of the foot. For a long time it was thought that this muscle was characteristic of man, but after anatomists began to look at it in some detail, it was found that probably in only about 85 percent of human bodies was it completely separated. In 5 percent it was not separated at all, and in 10 percent there was only a partial separation. In most gorillas this muscle does not exist because it is not separated from the long extensors. In muscles of the foot of some gorillas that have been dissected, however, this muscle has in fact been found. So gorillas share with man the possession of this unusual muscle, and neither the orangutan nor the chimpanzee has ever been shown to possess it. There are other muscles which show an anatomical relationship between man and the gorilla.

The gorilla has certainly developed in the direction of brawn. There are no fossil remains of any anthropoid that show the same degree of massive development as the gorilla. It exceeds man and all the other anthropoids in the weight of its body, the size of its lungs, the magnitude of its teeth, the massiveness of

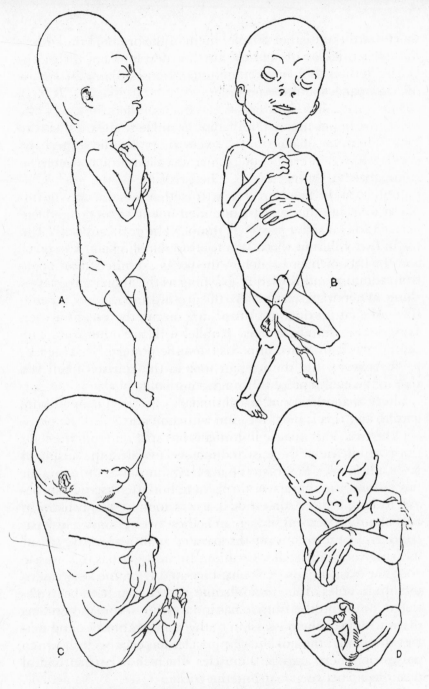

A and B, outline drawings of a human fœtus (said to be 4½ months old), from
photographs, and of the actual size of the specimen.

C and D, similar drawings (of actual size) of fœtus of gorilla.

Drawings of human and gorilla fetuses from W. L. H. Duckworth,
"Notes on *Gorilla Savagei*," *Journal of Anatomy and Physiology*,
Vol. 33 (1898), p. 82.

its jaw, and, of course, the strength of its muscles, but particularly the muscles of mastication. Despite all these facts, the young gorilla, at birth just under five pounds in weight, is smaller than that of a newborn child and is just as helpless. It has a smooth, rounded skull like that of a human infant, and the masticating muscles are no bigger than those of a human infant. Later on, of course, in the male gorilla the large bony crests develop on the skull to take up the enormous muscles which develop there.

Keith was very impressed with the fact that many gorilla characteristics resemble the glandular condition called acromegaly in man. In 1894, when acromegaly, a disorder of the pituitary gland, became recognized, Dr. Harry Campbell noticed that men and women who had this disease underwent a coarsening of features and other growth changes similar to those which occurred in the gorilla during its phases of maturation. At that time he explained acromegaly in man as an atavism or reversion to an ancestral or perhaps evolutionary condition. We know now, of course, that acromegaly is caused by an increase of the pituitary growth hormone. Keith believed that in the evolution of the gorilla the pituitary gland, and particularly that section of it concerned with the production of growth hormone, has played an important part.

Other items of anatomical interest are that when the gorilla grasps something, it grasps with the knuckles up as distinct from a human, who grasps objects with the palms up, using the fingers and the thumb. A gorilla's method of grasping is probably due to the fact that it does a certain amount of climbing, particularly when it is young, and this way of gripping with the hand enables it to contract the muscles of the upper arm to pull it up to the next branch.

In the bottom part of the leg there is a long thin bone called the fibula, and in the gorilla the fibula does not extend as far as the ankle, as it does in man, where it helps to prevent turning or twisting of the ankle. Other anatomical protections against the twisting of the ankle if the gorilla walked upright are also not present. The gorilla, therefore, does not extend its foot so that the sole is flat on the ground. In fact, the gorilla actually walks on the outside of its foot. The sole is always turned in. The gorilla also has a very small calf muscle, another factor which makes the adoption of erect posture very difficult. In man the calf muscle is attached to a long tendon which extends down to make contact with and be inserted into the heel bone, known as the os calcis, so that the bottom part of the leg develops a rather slender appearance. This part of the leg, together with the ankle, particularly in some young ladies, is a very attractive part of the anatomy. In the gorilla, however, the calf muscles extend all the way down with this tendon, which is known as the tendo achilles or the Achilles tendon, and inserts directly into the gorilla heel bone. This greatly thickens up the ankle and lower part of the leg so that it does not look as elegant as it does in humans and is almost certainly not the part of the lady gorilla that the gentleman gorilla pays special attention to. This thick ankle does, however, provide a great deal more individual strength. The Achilles tendon, as the famous myth explains, is a critical part of the human anatomy, and cutting or snapping or pulling the tendon away from its insertion in the heel bone is a serious matter for a human. For any of these to happen to a huge ape in the wild would be a disaster of the first magnitude, and this strengthening of the Achilles tendon by many muscle fibers which attach independently to the heel bone helps to prevent this kind of accident from occurring.

Some interesting studies were carried out at the Yerkes Primate Center by Dr. Russell Tuttle and Dr. John Basmajian, in which they inserted small electrodes into the muscles of the arms of a young gorilla, Inaki, a little female gorilla that was obtained from the National Zoo in Washington by swapping a young orangutan for her. This little gorilla was brought down to Atlanta by the keeper who had raised her from birth. She was clad in a diaper and wrapped in a baby's blanket and carried onto the plane. The keeper occupied a first-class seat on an Eastern Airlines jet and nursed the baby gorilla all the way to Atlanta. This unusual event attracted very little attention from the first-class passengers. One passenger looked up and asked what the animal was and was told it was a gorilla. He grunted and went back to reading his *Wall Street Journal*. The other passenger sitting next to the keeper with the gorilla said nothing at all the entire journey, and the only people who really were intrigued and took a great deal of notice of the little creature were the stewardesses. After some years in Atlanta, Inaki was chosen for studies of the muscles of the arms. The electrodes that had been inserted into the muscles were connected by wires to a machine known as an electromyograph, which measures the size of the electrical impulses generated in the muscles as they are used. They got Inaki to walk around in a room in a quadrupedal position, using all four limbs. Of course, as we know, when gorillas walk, they walk on the knuckles of their hands. A record was made of what muscles were being used when she did this and also when she was swinging on a trapeze. These electromyographic studies on the arm muscles of the gorilla showed that the elbow joint was apparently especially adapted both for knuckle walking and for suspensory behavior (swinging from trees). Probably her ancestors spent a long period of time in the trees before they became semiterrestrial animals and got around on their knuckles. These electrical changes which occur in the muscles of the gorilla are very different from those in man. Man and gorilla presumably shared a common heritage in adapting to a tree-living existence, which involved vertical climbing and hauling themselves up with their arms and swinging by their hands, but eventually in man the muscles involving powerful and rapid extension of the elbow joint became very important and this enabled the hands to manipulate tools accurately, a character which helps to distinguish him from the apes.

We have talked about the mountain gorilla and the lowland gorilla, but Dr. Colin Groves has distinguished a third subspecies of gorilla. This is the eastern lowland gorilla found in the east, but not in the mountains.

It is intermediate in some respects between the true mountain and the true lowland gorillas. It is not as black as the mountain gorilla, but its hair is longer and more silky in texture. In the proportions of its body, however, it is more like the lowland gorilla. The face is long and narrow, and this makes possible immediate distinction from the other two gorilla races. The lowland gorilla can be distinguished from the mountain variety by a small "lip" of skin in the former, which overhangs the top of the nasal septum. A very small "lip" of this type may also be seen in some eastern lowland gorillas, but mostly it is absent.

While there are a few slight anatomical differences among the three gorilla races, there are no observable differences among the brains, and in fact as far as structure is concerned, there is little difference except in size between the gorilla and the human brain. Even the size difference is not obvious at birth, the brain of the newborn gorilla being identical in size and shape with that of the human baby. After birth, however, the human brain continues to grow rapidly, whereas that of the gorilla grows slowly and soon gets left behind, though they remain similar in structure and appearance; in fact the brain of the gorilla and human coincide so well that if it weren't for the difference in size between them, they could easily be mistaken for each other.

There is a part of the brain called the olivary nucleus, which is related to the regulation of the movements of the eyes, the head, and the hand and their integration with one another. This nucleus is essential for the ability to acquire tactile skill—that is, skill with the hands. This nucleus is bigger in the gorilla than in any of the other apes. Despite this fact, however, we are still not quite sure whether the gorilla does show greater manual dexterity than other apes.

In the human most of the skull is concerned with housing the brain. In the gorilla a much larger part of the skull is devoted to forming the face and the snout and also for providing attachments for the chewing muscles and the neck muscles. Many people have compared the weight of the brain with the weight of the body and produced an index presumed to give some indication of intelligence, but this does not work because many small mammals have brains which, compared with their body weight, are quite large. The gorilla has a much smaller brain than man. He also has a much heavier body.

Among the fossil remains believed to represent a possible ancestor of man is *Australopithecus*, discovered by Professor Raymond Dart in Africa. It had a brain which has been estimated to have weighed about 450 grams. This is about the same size as the gorilla brain, which suggests that the *Australopithecus* may not have been any more intelligent than a gorilla. But if one compares the weight of the body with the brain, *Australopithecus* is way ahead of the gorilla because he had such a small and light body. One of the estimates makes the body weight around 70 pounds. So a 450-gram brain in a 70-pound body provides a very good ratio when compared with the gorilla's 450-gram brain in a 350-pound body. *Australopithecus*, however, still has a lot of its skull associated with the face and is very unhuman from this point of view. Man's body is similar to that of the African anthropoid apes, that is to say, to the gorilla and the chimpanzee. There is a similarity, too, in the blood, which also suggests he is a close neighbor to these two apes. If we assume that a common ancestor gave rise to both the apes and to man, what would cause one type of ancestor to become a gorilla or a chimpanzee and the other to become a human being? The ability to stand permanently erect on two legs and run on two legs, known as bipedalism, is a major distinguishing mark between apes and man. Frank B. Livingstone of the University of

Michigan has considered this particular problem. Perhaps the concentration of the ancestral animal or animals in a geographically isolated area was the key to the acquisition of bipedalism. There were two activities in Africa during the Pliocene period (14,000,000 years ago) which probably accelerated the change from apelike ancestor to man by forcing him onto his two legs. These two were first, the spread of the Kalahari Desert, which would geographically isolate some areas, and second, the seismic activity in the area of the Rift Valley, west of Kenya. As a result of these phenomena, apelike ancestors were probably cut off in isolated areas. The reason that isolation would be important is that it would produce what are known as contact lines between the forest and other areas, and to penetrate into areas beyond the forest the animal would need to develop new types of locomotion and other behavior. In other words, the animal that was going to turn into the human was associated with the edge of the tropical forest and did not live deep in the forest, so that from the edge of the forest he could make forays out into savannalike country or, in the Rift Valley area, could leave the forest and climb into mountainous areas where the vegetation was very different.

The present-day gorilla is what is called edge specific, that is, he prefers the edge of the tropical forest. For instance, not only are the animals found on the edge of the forest where there is some savanna country, but also they occupy areas where there is a mountain area, and they will spread along rivers that run through a tropical forest. This also constitutes a fringe habitat. The gorilla is predominantly a terrestrial animal, and its foot is anatomically much closer to the human foot than to that of any other ape. Also since the gorilla is very large and even the moderate-size animals do not spend a great deal of time, and the bigger ones no time at all, in the trees, the animal is able to defend itself against other animals by brute force rather than save itself by climbing up into the trees as perhaps a chimpanzee might do.

Although the gorilla normally proceeds on all fours, there are times when it does rise onto its two feet. For example, it likes to thump its chest with its hands, as we have indicated elsewhere, and it stands on its hind legs to do so, so that it becomes bipedal at this time. Furthermore, at the end of its charge it also tends to be bipedal. Gorillas will charge over long distances, running on all fours, but when they come just to the point of making the attack, they rise up on their legs to do it. Although a gorilla will bite when attacking, more effective than this and more commonly used are the very powerful arms, which pack enough strength to kill an animal even with a passing swipe and are certainly powerful enough to kill a human being.

Sherry Washburn, writing in 1960, suggested that the *Australopithecus* was also probably not able to stand erect for any length of time and that he would run or walk for short distances and then squat and then run and walk and squat again. This is in many respects similar to the locomotion of the gorilla. The implications of all this presumably are that in animals that occupied this forest-edge situation there were some that became completely bipedal and eventually turned into an *Australopithecus* ancestor and then into a human, whereas another branch of the family now represented by the gorilla, while developing the ability to be slightly bipedal on occasions, never really made it through to the human type. What the factors were which propelled the gorilla and human ancestors along their respective paths it is difficult to say. Presumably in one group there was a series of mutations which helped to favor the development. Of course, once an animal became bipedal it was able to carry food and manipulate sticks and stones and objects to protect itself.

Although Dr. Yerkes' studies on the mountain gorilla Congo suggested that the animal had mental limitations, other studies have demonstrated that gorillas are at least as intelligent as the chimpanzee, but their intelligence is in a slightly different direction from that of that particular ape. Gorillas, for instance, have difficulties in solving problems when implements are involved. Chimpanzees, on the other hand, even in the wild, use sticks to obtain food. For instance, they will put a stick into an ant hole, wait for it to be covered with ants, and then pull it out and suck the ants from it. They even squeeze leaves and use them as sponges to extract rainwater from holes in trees or tree stumps. Gorillas have never been observed to do either of these two things in the wild, and they do not seem to be able to use a stick to rake in food in the way a chimpanzee does or even to use it in knocking down fruit which is suspended out of their reach.

Charles Cordier, who spent a good many years in the Congo catching animals of various sorts for zoos, was once charged by a large gorilla. This animal held a long pole in front of his chest, but did not project it in any way at Cordier, but there are descriptions of young male gorillas breaking off the branches of trees and throwing them at hunters. Gorillas seem to be very much more deliberate than chimpanzees, particularly in their approach to tests which presume to appraise their intelligence. They will, however, concentrate on a problem and are very persistent in their approach to it and in some cases seem to be more interested in actually solving the problem than in getting the food reward at the end of it. They also seem to have a better memory than chimpanzees and seem to discriminate among various geometric shapes more effectively than chimpanzees do. Gorillas can be very gentle, and there are many examples of their being very loving. The young, particularly, need affection and if captured too early and not given the care, attention, love, and affection that they normally get, they are likely to die. They also have many human characteristics such as laughing and crying and throwing temper tantrums.

Dr. Ray Carpenter noted that two captive gorillas that he studied at the San Diego Zoo drank large quantities of water, as much as two gallons a day. Normally they put their mouths to the water and suck it in. When they were excited, however, during periods, for instance, when they were playing, they used their hands to dip the water up in the air higher than their heads and then let the water fall out of their hands and into their mouths. Sometimes they took various containers and dipped out the water and drank it from the container. This is particularly interesting in view of the fact that they are not seen to drink at all in the wild state. He found, and this is something that we have also found at the Yerkes Center, that they are very casual about excretion from the body, particularly defecation. They never take up a special position to perform this act, but it occurs while they are walking around, while they are playing, or while they are lying down. Urination, however, was different, for the apes stopped and usually stood in a particular position while they urinated.

Gorillas have various types of play. They will play with themselves; they will use manipulatory play. They will run or roll or spin and, of course, beat their chests from time to time. They play very actively with different objects, for example, ropes, balls, sticks, water, tires, decorative objects, and even small live animals if they have access to them. They also perform typical social play such as wrestling, chasing, teasing. Their play in general is vigorous and persistent and varied, but it does not seem to include play fighting. They play actively with straw and may carry it under their chins. They will also grasp vertical ropes and spin on them. Gorillas are also fascinated by water running from a hose. If they are given the opportunity, they will find the end of the hose and then open their mouths and direct the water from the hose into their mouths. Two animals that Carpenter examined competed for this opportunity as much as they competed for food.

Gorillas also like to slide. Gorillas at the Yerkes Center do this continuously if the cement floor of their cage is wet. They run and slide from one end of the cage to the other. The gorillas that Carpenter examined would even flush or dash water from their pool onto the cement, so that a skating surface could be formed. His animals skated with their arms and hands. Yerkes animals skate on their knuckles and feet. Carpenter gave the two gorillas he was studying an inflated rubber ball, and they played with it and tried to take it away from each other until it burst. When it burst, the animal that had been holding it seemed to be very puzzled, even though he still held the remains of the rubber in his hands, and looked around to find the ball with which he had been playing. He could not understand what had happened to it.

Chasing play is also fun for gorillas, and it tends to mix with a certain amount of wrestling. Sometimes the animals carry out intermittent wrestling for a number of hours, but the actual periods of vigorous personal contact do not last more than a few minutes, sometimes only a few seconds. Chest beating is also common among gorillas in captivity. When gorillas slap their chests, they do it with a rapid alternate slapping with the open hands against the upper part of the chest. The animals will also beat logs and floors in a similar manner during play situations. When gorillas make the characteristic chest-beating rhythm, they are usually in a playful and rather contented mood.

Grooming is characteristic of subhuman primates, and from that behavior has probably descended the affectionate and petting behavior carried out by humans. Grooming in gorillas seems to occur rarely, and it is different from the pattern of grooming that is found in chimpanzees. Certainly it occurs a lot less. There are two types of gorilla grooming. There is self-grooming and there is social grooming. The activity is pretty much the same in both cases. It includes parting the hairs of the coat and fingering the skin that is thus exposed. Both hands are used, but the right hand in most animals seems to be the most active. Any particles or scabs that can be seen are removed by the index finger and thumb. Sometimes the lips are used to remove structures on the skin. It is obvious that this behavior is extremely important in the removal of ectoparasites such as ticks from the body. The animals often lie on their stomachs and have their backs groomed or lie on their backs and have the abdomen and chest groomed. The abdomen is groomed more frequently than the upper part of the chest, perhaps because the skin is softer or there are more ectoparasites in that region. Sometimes apes get small cuts and bruises, some of them serious and some not. They give detailed care to such abrasions. They have been seen to suck an injured hand or hold it in water. Sometimes they pull hair from the area which has been wounded and sometimes simply touch or lick the broken skin. If the injury is in a part where the animal cannot get to it himself, this kind of treatment is often given by another gorilla. It is not possible to tell whether this treatment helps the healing, but at least it does not seem to retard it much because most cuts and bruises seem to heal within a few days.

11

What to Say to a Talking Ape

THERE have been many speculations as to why monkeys and apes don't talk, and the suggestion has been offered in some circles that they are too smart to talk because they know that if they talk, they will be put to work. Now that chimpanzees and gorillas are being taught to communicate with humans, this level of reasoning may be upset.

Human infants make noises from the moment they are born, and these sounds are called crying. They differ in this respect from infant apes, which do not cry but do make a certain modest whimpering sound. The development of speech in the human infant starts with a cry which develops gradually to a babble and finally to the acquisition of words. In the neonatal period the human infant is limited in the number of sounds it can make, but there are various types of cries which have been listed: birth cries, fussing cries, gurgles, hunger cries, shrieks, and inspiratory whistles. Most of these cries can be made spontaneously. The limitations on the types of cry that the neonate can make is due to the structure of its vocal tract, which makes it impossible for the infant to carry out vocal maneuvers characteristic of the human adult. The shape of the vocal tract in the human infant has a uniform cross section, and it is similar to that of the adult great apes—more so in fact than it is to the human vocal tract. Here, perhaps, is the secret why the gorilla, chimpanzee, and orangutan are not able to use human speech. Similarities between the human baby's vocal tract and that of the great apes are the following: First, the larynx, or voice box, is quite high in the human baby and is located in a position similar to that which has been described in the gorilla. Also the newborn infant's tongue is large by comparison with the oral cavity and virtually fills it. This means that the use of the tongue for vocal manipulation is limited in this early period. Later in development the lower jaw enlarges so that the tongue does not completely fill it and can become a great deal more mobile. Mediation of speech is affected by a number of factors. One of them is the ability for the root of the tongue to move and constrict the size of the vocal tract. Of course, human infants, unlike monkeys and apes, eventually are able to produce a complete range of human speech. Monkeys and apes have never got beyond the human neonatal condition. The apes, including the gorilla, also lack a fixed tongue root which would be able to form a movable anterior or front wall to the larynx. Nonhuman primates have long, flat tongues, and the tongues appear unable to contribute to the changes in what is described as the supralingual vocal-tract shapes that are necessary for human speech.

Dr. Philip H. Leiberman, who has studied this subject in great detail, points out that the human pharynx varies as much as ten times in the production of the vowel sounds *a* and *i*. The vocal tracts of apes, by comparison, have the resonating characteristics of a tube which is more or less uniform; although it may be expanded at the end in a trumpetlike fashion, the animals are unable to vary the cross section of the lumen in the way the human does. This is necessary for the production of human speech. The study of the remains of Neanderthal man has also suggested that his vocal tract was inadequate for the reproduction of the vowels *a, i,* and *u,* so he probably was not able to use speech in the sense that we know it today. He probably made a great variety of noises and accompanied them by a good deal of gesticulation. It has been suggested that before the development of true speech, man tried a crude communicating system consisting of gestures accompanied by cries and grunts. It has been theorized that these could have lasted for hundreds of thousands of years. The Aurignacian period, 30,000 years ago, was probably when the development of human speech as we know it began. There was obviously some form of speech, however, which existed for some thousands of years before this, but it was almost certainly not the type of speech we have today.

In the absence of sophisticated vocal communication, apes such as the gorilla require a more extensive visual means of communication. Chimpanzees use many gestures, and Drs. Beatrice and Allen Gardner of the University of Nevada, working with a chimpanzee, Washoe, in an attempt to communicate with her, decided to use gesturing procedures because the early work of Dr. and Mrs. Keith Hayes at the Yerkes Center with chimpanzee Vicki showed that vocal communication was limited. Generally chimpanzees vocalize when they are excited. Gorillas, particularly male gorillas, as we have seen, especially in the wild, are capable of making a powerful and devastating roar.

The use of the chimpanzee hand in communication is prominent. For example, at the Yerkes Center chimpanzees, particularly the young ones, commonly use the begging gesture spontaneously. The hand is directed toward the human, and with the palm up asking and with a beckoning gesture, you are supplicated to put food into it. The films taken at the Yerkes laboratories over the years demonstrate a variety of types of manual gesturing by chimpanzees, and quite striking is the one used to gain the cooperation of another chimpanzee. This simply consists of taking the animal by the hand and leading it in the desired direction. A soliciting chimpanzee may also put one arm around the other chimpanzee's neck and try to lead it by this method or even to wave it in the direction in which the other chimpanzee wants it to go.

Although the chimpanzee does not seem to be able to speak the human language, it does seem to understand a large number of commands and words said to it by the human being. We know of one chimpanzee which had no trouble in understanding forty words of command and could make the appropriate and correct response to these commands every time.

A number of scientists have attempted to communicate with chimpanzees by using other methods. The animal chosen in the first instance for this purpose was the chimpanzee Washoe, which was used by the Gardners in their experiments. The gorilla and the orangutan, more placid and withdrawn animals, have not been used except recently because they were not highly motivated. Presumably they do not think humans are worth communicating with. The Gardners therefore chose to communicate with Washoe with gestures. There are two types of communication used by deaf people which they could utilize. One of these involved the use of a manual alphabet. In other words, the position of the hands and the fingers correspond to the letters of the alphabet, and words are spelled out. This is the more commonly used type of communication among deaf people. There is another system, the American Sign Language for the deaf. This system consists of a number of manual configurations coupled with gestures that are equivalent to specific words or specific concepts. Two examples of the American Sign Language are presented by the Gardners in a recent publication. One of these is the sign for "always," in which the hand is clenched with one finger extended and the arm is rotated at the elbow. In the sign for "flower" the fingers are all extended and touched first to one nostril and then to the other, as if a flower were being sniffed.

The babbling of human infants has been suggested as the origin of language, and it is shaped eventually into language by adults who rear babies. Similarly, young chimpanzees, especially if raised in association with humans, undergo what might be described as manual babbling. Its transformation into the elements of language is very well described by the Gardners:

> We encouraged Washoe's manual babbling by being as responsive as possible. We would clap, smile, and repeat the gesture that she seemed to have made, such as people might repeat "goo goo" in response to the vocal babbling of the human infant. When she babbled a gesture resembling a particular sign of ASL [American Sign Language], we would try to initiate some appropriate activity. For example, during the period when manual babbling was common, Washoe was fond of touching her nose or her friend's nose with her index finger. This is very similar to the ASL sign "funny" in which the extended index and second fingers are brushed against the side of the nose. We could make a regular nose touching game of this; sometimes she initiated a game and sometimes we did. Everybody laughed as though it were all very funny. At some point Washoe introduced a variation which consisted of snorting if the nose was touched. It was a simple step to initiate the nose touching game whenever something happened which might seem funny to an infant chimpanzee. Gradually, Washoe came to make the funny sign in funny situations without any prompting.

In an article published in 1971 the Gardners listed eighty-five words that the animal had learned in American Sign Language. These included "come," "gimme," "more," "up," "sweet," "open," "tickle," "go out," "hurry," "listen," "toothbrush," "drink," "hurt," "sorry," "funny," "please," "food," "flower," "cover," "dog," "you," "bib," "in," "brush," "hat," "knee," "shoes," "Roger," "smell," "good," "Washoe," "pants," "clothes," and "cat." Some examples of the types of gestures she has to make for some of these are as follows. For "listen" she closes her hand, extends the index finger, and touches her ear. For "toothbrush" the index finger is extended from a closed fist, and the side of her finger is then brushed back and forth across the upper teeth. For "drink" she extends the thumb from a closed fist and touches it to her lips. For "hurt" (pain) she takes the two fingers from two closed fists and touches the fingertips at the side of the injury. Washoe began to combine signs after ten months, when she was between eighteen and twenty-four months old; this is quite close to the age when human infants begin to combine words into two-word combinations. The two combinations she communicated were "gimme sweet" and "come open." Washoe learned to use these signs for various actions and people in the correct association. The Gardners give a particular illustration for the signs she used for "Greg," for "Naomi," for "hug," and for "tickle." "Hug" and "tickle" were signs that Washoe used for action, and Greg and Naomi were two of the helpers who were obviously capable of hugging and tickling. Washoe was able to form the concept that Greg was Greg and was called Greg whether he was hugging or tickling. Therefore, she was able to use a combination "Greg tickle" or "Naomi hug," but did not use meaningless combinations such as "Greg Naomi" or "tickle hug."

In the training of young children the child will often use one word or a two-word combination, and the mother will then expand on it. The Gardners give an example of an actual conversation between a child and a mother in which the child says, "More cookie," and the mother says, "Your cookie is right here on the table." The child says, "Cookie," and the mother says, "No more cookies. Later we will have a cookie." They described how they used this technique in expanding what the child says in the training of Washoe:

> A good example of Washoe signing out [is] when [she was] near the door of her house. Persons familiar with Washoe understood this as, "I want you and me to go outside." Often we granted her request, but sometimes we denied it, signing a simple "no" or elaborating the denial by explaining that it was too cold, or too dark, or too close to suppertime. Washoe's response to the denial ranged from desisting, through persisting by signing again, to emotional outbursts such as whimpers or temper tantrums. If we tested the interpretation of "out" by partial fulfillment of the request such as thrusting her out by herself, or going out without her, we were certain to illicit

strenuous resistance and the emotional outburst. We remained confident of this interpretation even though early combinations for the situation were of the form "please out" and "hurry out." Washoe began to use the two pronouns "you" and "me" in January, 1968 (month 19 of the project). By Spring, 1968, she had signed "you me out" in the doorway situation and later produced many variants such as "you Roger Washoe out," or "you me go out" or "you me go out hurry." When Washoe used one or more signs one after the other, they could be normally interpreted as a combination, but sometimes there were periods of various length between the individual signs, and it was almost impossible to be certain that they were meant to be a combination. It was therefore essential that she have included in her vocabulary such joining words as "and" "for" "with" "to" when she was in her 36th month of training.

The Gardners point out that this type of word is also not present in the early sentences of children. It became necessary to distinguish when Washoe's communication had finished by noting the rest position to which she returned her hands when she had finished making a communication.

Another pair of researchers, David and Ann Premack, also attempted to teach language to a chimpanzee called Sarah. They used, instead of a sign language, a number of pieces of plastic of different shapes and colors, each representing a word. These pieces of plastic were backed with metal and were used with a magnetic board so that they would stick to it in any position required. Dr. Premack and his wife succeeded in teaching Sarah a vocabulary of 130 terms at least. They say she uses them with a reliability of between 75 percent and 80 percent. They asked the question in an article in 1972 in *Scientific American*, "Why try to teach the human language to an ape?" They felt that in their case the motive was to try to discover the fundamental nature of language. They pointed out that languages are said to be unique to the human species, and it is well known that other animals have fairly elaborate communications systems, although it is doubtful that they actually have anything equivalent to human language. Human language is a highly refined form of communication, although it is possible that some aspects of it which are thought to be unique to the human actually belong to a more general system of communication. The Premacks believe that if "an ape could be taught the rudiments of human language, it would clarify the dividing line between the general system and the human one." They obtained a good deal of success with Sarah and are now working with an orangutan which was born at the Yerkes Center and which was sent them on loan for this purpose.

The Premacks have undoubtedly been successful in teaching a simple language to their chimpanzee, Sarah, and it has been done in such a way that a complex subject has been reduced to a series of simple steps which were highly accessible to the chimpanzee. Although the experiment was not specifically designed for any other purpose, it does in fact have some important spin-offs since the program that was designed to teach Sarah to communicate has now been successfully used with people who have language difficulties caused by brain damage. It could also be used with advantage in teaching the autistic child. Here is yet another example of how a fundamental study carried out without any real practical intent will eventually nearly always have some kind of practical application.

The Yerkes Center study on language with the great apes, which is under the supervision of Dr. Duane Rumbaugh (of Georgia State University) and Professor Harold Warner (of the Yerkes Center), with the assitance of Dr. Ernst von Glasersfeld and Dr. Leo Pisani from the University of Georgia, is being carried out primarily on a chimpanzee, Lana. In the beginning it was planned to use an orangutan also, but he was so frightened of the equipment and fell so far behind the chimpanzee that it was necessary to take him out of the experiment. He would obviously have to be trained under different circumstances.

The Yerkes Center has not used a gorilla in any of these language activities yet, but it intends to use one later. In the case of the Yerkes project, it has been planned that communications between the chimpanzee and the human take place with a computer for the intermediary. This eliminates any possible human error in the communication of the chimpanzee, in other words, interpretation by the human or the human giving some type of cue other than the planned gesture or cue.

In the Yerkes experiment there are a series of consoles with lighted square keys which are about 1¼ inches by 1 inch. They burn at half brilliance when the machine is on, and whenever one is pressed, they light up at full brilliance. There are five banks of keys; the first bank contains agent words and pronouns, the second bank activity words, the third bank prepositions, the fourth bank object words, and the fifth bank other words such as "yes," "no," "not," etc. Initially the animal was asked simply to request things by a single word. For instance, if she wanted a piece of apple, she simply depressed the key which had the sign for apple on it. Later on she was asked to attach words so that she now says, "Please, machine, give apple" or "Please, Tim, come into room" or "Please take Lana out of room." She has also developed to the degree now that she combines sentences. She also asks for the names of objects which she does not know. Lana can ask for a variety of foods and drinks. She can also ask for toys, all of which are delivered by the computer. She can also ask the computer to show her a movie or lantern slides or to play her music. She often asks for the last two or three when she is on her own in the evening when everyone has gone home. There is no doubt that the development of language with Lana is opening up a very important insight into the chimpanzee mind.

There are plans to try this technique with the gorilla, and we anticipate that the gorilla will do well in this project, particularly in view of the studies which are already being carried out at Stanford Research Institute in California in association with the San Francisco Zoo. Penny Patterson, who is a graduate student in developmental psychology, is attempting to teach one of the zoo gorillas, Coco, the American Sign Language for the deaf, using the technique which has been used by the Gardners.

Penny Patterson and gorilla Coco from the San Francisco Zoo. Coco gives sign for "viewfinder" with his right hand. (Courtesy of Jim Ellis and Oklahoma City Zoo.)

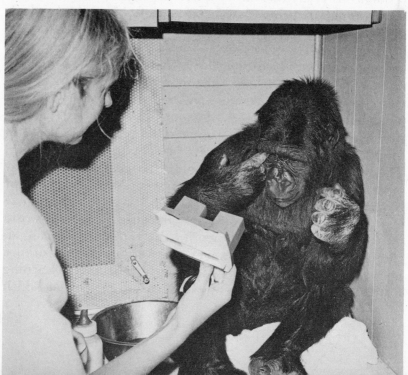

Penny Patterson and gorilla Coco. Coco signs "Me up," asking to be picked up.

Coco, three years old, is a female lowland gorilla. Her place of residence is really the San Francisco Zoo. She is isolated from other gorillas by being placed in a comfortable trailer in which the experiments are carried out. Penny Patterson has been working with this animal now for a year. Coco has been able to communicate with her using the ASL for the deaf to request a number of personal needs and, according to a news release, to express her personal feelings and satisfy her curiosity. She has a vocabulary at the moment of about 140 words, and the signing conversation she makes appears to be self-initiated and spontaneous; she uses about sixty to eighty signings each day. The words she uses most often, of course, relate either to play or to food and may include such phrases as "come tickle me" or "come swing me" or "chase me." She also uses sign language to express a desire to go out of the trailer into the outside world for a period. She specifies types of foods that she may want, including bananas, apples, oranges, and various types of drinks. She can ask for different toys. The rate at which she is learning words and combinations of words seems to match that which the chimpanzees can do. Plans are made for Coco to continue her sign-language studies until she reaches the age of maturity, which of course will be around eight years of age.

Ms. Patterson said that according to a newspaper reporter in the Atlanta *Journal-Constitution,* it won't be a lonely coed's life for Coco. Kong, a gorilla that lives at Marine World-Africa, U.S.A., in San Francisco, would come to the campus every now and again to date Coco. Ms. Patterson said that if the two hit it off, other arrangements could be made later. Coco was born at the San Francisco Zoo on July 4, 1971, and was the daughter of Bwana and Jacqueline. Bwana, the lord and master of the zoo's gorilla harem, according to a news release, came from his native West Africa by airplane in the arms of a veterinarian as a baby in 1959. Jacqueline was purchased from the Brookfield in Chicago by Carol Soo-Hoo in 1964 as a gift to the San Francisco Zoo. Coco is now around sixty pounds and of course is quite a strong animal. Coco has a number of toys to entertain her. These include a swing, a tire suspended from a rope, dolls, a toy train, a toy truck, Playskool equipment, and stuffed animals. She enjoys playing with her toys very much, but she also spends some time observing the other gorillas in their activities when she is out in public. It appears, therefore, that the gorilla, in keeping at least with the chimpanzee and probably with the orangutan, has an ability to communicate in quite a complex form with humans and perhaps even in something that might be called language. This is an extremely important development and should affect our whole thinking about the way great apes are kept in zoos as if they were in a penitentiary. It accents the desirability of animals such as the great apes being given something to occupy them mentally when in captivity to save them from getting bored and to make them happier animals.

Gorillas understand a lot of what goes on around them. Like all great apes, they can look at a photograph and understand it. We know this from the studies made with our apes by Dr. Richard Davenport on what is called "cross-modal transfer of information." In this type of experiment an animal is shown an object and he can then pick it out from a number of other objects without seeing it, using the sense of touch only. Furthermore, he can pick the object out even if he has seen only a photograph of it. The photograph can be of the soft type, which shows a lot of detail, or a hard one, which shows only the form of the object; it is all the same to the animal. This ability to transfer information from one sensory modality to another is believed to be essential in human children if they are to learn to read. Since gorillas and other apes can understand pictures, they might be expected to understand television and movies, and there is a good deal of evidence that they at least appreciate it even if they don't understand it. Our communicating chimpanzee, Lana, likes to see a movie which we have called *The Developmental Biology of the Chimpanzee.* By tapping out on her console, "Please, machine, make movie," she can see it whenever she likes. So far she has asked for it more than 200 times. She does not understand what it is all about, but she almost certainly recognizes the baby and juvenile chimpanzees in it.

Some years ago we acquired from Mrs. Elizabeth Borgese, one of the editors of the *Encyclopaedia Britannica,* a little chimpanzee called Bobby. Bobby had been brought up in Santa Barbara and had been introduced to television; he proved to be very fond of Western movies and soon learned to pick out the bad guy because whenever he appeared on the screen, he was associated with violence. In fact, Bobby used to get excited whenever he appeared on the screen even before the violence started. After we received Bobby, we thought we would try the effects of TV on our other apes. Sometimes for the purposes of an experiment an animal had to be isolated in a room for a time and she would get awfully lonely, so we thought that a television set in the room would be just what she wanted, and so it turned out to be. Sometimes a nursery youngster would get a cold and have to be isolated so that he would not infect the others. A television set in the room with him made all the difference. Fortunately the Sylvania Company made us a present of a dozen used TV sets and that helped us along.

Our use of TV to keep our chimpanzees happy was written up in a very amusing fashion by Neil Hickey in *TV Guide* of October 3, 1970.

The experiences with TV at the Yerkes Center stimulated Jack Throp, the director of the zoo in Honolulu, to get the local TV station, KGMB, to donate a seventeen-inch Sony portable television set to the zoo. It was one that the director had used in his office, but he gave it up for his gorillas, Congo and Cameroon, to use. It had been put into a special Plexiglas, gorilla-proof container and was put by the gorilla cage, but far enough away from it that they couldn't reach it, although of course there is a possibility that they could throw things at it. Another activity that the zoo carried out to give a little more variety to life for the gorillas was to change them every three days from a cage which faced toward the sea to a cage which faced toward the mountains, so they had changing views twice a week. Incidentally, it had the advantage in giving the zoo keepers a chance to clean one of their cages thoroughly after they had vacated it. They were also given piles of palm leaves, coconuts, tires, and big pasteboard boxes to play with. The director of the zoo said that his gorillas were great people watchers, but that they must get terribly tired of that after a time. When the gorillas had the opportunity to see the *I Love Lucy* show on television, Cameroon first of all threw a handful of water at the plastic screen which protects the front of the television set. Then he threw a handful of feces, which the embarrassed keeper hastily washed off the screen. While Cameroon, who was twelve years old and weighed 400 pounds, acted up, Congo just peered at the set. He got an unusual view of the program, at least in the beginning, because he watched it hanging upside down from a pipe which is located near the top of his cage.

12

Everything You Always Wanted to Know About the Sex Life of the Gorilla, But Were Afraid to Look

MAN may be the sexiest of all primates, and the gorilla may well be the least sexy. This certainly must come as a surprise to those who believed the gorilla to be the supersexed stud of the animal kingdom, but experts give a very low score to the gorilla as a lover. George Schaller, who spent more than eighteen months in the wild observing gorillas, witnessed only two copulations; in fact, these big apes carry out their sexual activities in a most casual manner. Another sexual myth which has recently been shattered had this so-called rapist ape endowed with an organ of inordinate proportions. The truth of the matter is that the gorilla's penis is extremely short in relationship to his body size. Swiss zoo director Ernst Lang estimates five centimeters, about two inches, for the length of the erect penis in an adult male. One gorilla in the Milwaukee Zoo had a penis of only 2.5 centimeters while it was masturbating. One of the longest recorded, measured at nine centimeters, belonged to an adult gorilla at the London Zoo.

The sexually indifferent gorilla is prone to copulate most often when the female is in heat, which is demonstrated by a slight turning out of the lips of the labia so that the pink interior flesh is seen. This is the time when the hormone levels are highest in the female; consequently they are more receptive to the males and will often extend the invitation to copulate.

Copulation between gorillas may take place in several different positions, which we will discuss later. The coupling procedure is a longer one than that of the chimpanzee, which is only a matter of seconds, but not as long as that of the orangutan, who is the sexual athlete of the primate world. His amatory exercises while hanging from a tree often last an hour or more.

When a captive female gorilla is feeling sexy, she often presents and rubs against the wall or the partner or even against another female if she is caged with her. Gorillas have picked up a technique of back riding just before or just after heat. Each sex takes turns in riding the other. It is hard to say if it is sexual play or just fun since it also occurs between animals of the same sex. Animals also drink each other's urine on occasions, but it is hard to guess what motivates them to that activity. Dr. Yerkes recorded that a female mountain gorilla that was only five or six months old mounted and thrust against a dog, and another time she threw herself on her back and pressed her genitalia against Dr. Yerkes' feet and tried to pull him down on her. She also presented to him on all fours. Dr. Schaller, in addition to his observations of copulation in the wild, has also described a female gorilla at the Columbus Zoo who backed against the bars separating her from a male, who was unable to carry on intercourse through the bars but did carry on what might be described as digital intercourse. He has also described how a mountain gorilla female in the Bronx Zoo ". . . made advances to the male in her cage by rubbing her rump against him, by holding his hand pressed against her genital area, occasionally turning on her back with legs widely spread. She fondled and licked the testes of the male. The male, in turn, rubbed the breast and the genital area of the female with his hand."

In the National Zoo in Washington, in which a number of gorillas have been born, copulation between gorillas has been observed in which the male lies supine with the female straddled and crouched down on top of him; copulation took place in this position with most of the pelvic thrusts being supplied by the female. On another occasion the animals were observed to copulate standing up with their abdomens together. Copulation has also been seen with the male in the dorsal position. The position most frequently observed, however, was the female crouching with her knees bent under her belly and her chest almost in contact with the floor. The male squats behind her, sometimes closing his hands on her hips, and enters from the rear. Dr. John T. Emlen, when observing gorillas in the National Zoo in Washington, saw the male thrusting very hard and rhythmically during copulation and after thirty to forty seconds giving a soft but rather rhythmical hooting duet with the female. The total copulation was only one to one and a half minutes long.

In the wild three precopulatory activities of gorillas have been described: (1) the starting *walk:* The male approached the female with short, abrupt steps. His body was held very stiff and erect, and his head was tipped slightly upward and sideways. The male advanced in this way toward the female and sometimes circled her. He always averted his head, but watched her from the corner of his eye. When approached closely, the female likewise turned her head to the side. (2) *Wrestling:* The male and female lumbered toward each other bipedally, slowly swinging their arms overhead until close enough to grasp each other behind head and back, a behavior also noted in chimpanzees by Dr. Harold Bingham in 1928. Bent at the waist, they pulled and pushed and grasped at each other's legs until finally they tumbled over. Then they sat and faced each other, mouths wide open, and emitted almost continuous deep growls and grunts. They rocked at each other's shoulders, and they hugged. Several times the male cradled the female in his arms. The whole sequence of events was slow and gentle and appeared ritualistic. (3) *Running:* The wrestling was several times interrupted when either the male or the female broke loose or left off on a march or ran around the cage with the other following closely. Circling of the cage has also been observed by Dr. Yerkes, in one of his articles in 1943, in sexually aroused chimpanzees, which in addition exhibit romping, teasing, petting, eating, fighting, and tantrums.

In the Basel Zoo director Ernst Lang has seen gorillas copulating face to face and also with the male behind the female. Dr. Schaller collected a good deal of data concerning copulation in gorillas and drew attention to the fact that the erect penis can easily be seen in these animals. If the male is sufficiently aroused, the vagina is penetrated quickly, but the animal thrusts only a few times and has an orgasm in about fifteen seconds. During the orgasm the male's body becomes very rigid. After the male has reached orgasm, the male and female gorilla may continue to fondle each other with some gentleness, which is a contrast to the behavior of chimpanzees.

A graphic account of gorilla copulation in the wild has been given by George Schaller. While up a tree observing a group of gorillas, he heard a peculiar sound, and after looking around for a time, he became aware that a gorilla called DJ was copulating with a female on the steep slope. She was on her knees and elbows, and he had mounted her from behind and was holding onto her hips. Since they were both on a steep slope, every time he thrust they slid downhill. Within about fifteen minutes they had slid forty feet and stopped sliding only when they finished up against the tree trunk. Then Schaller says: "A hoarse trembling sound, almost a roar, escaped from DJ's parted lips, interrupted by sharp intakes of breath." Apparently he had completed his orgasm and the female lay there without moving for about ten seconds before she got up and walked uphill, while the male stayed behind still panting. With the gorilla's attention on sex, Schaller was able to descend the tree and go about his business. He was also able to witness copulation on another occasion. This time it was initiated by the female. The Outsider, the big silver-back male who was not really a member of the group, but skimmed the periphery, seemed to be looking into the distance when a female gorilla appeared behind him, put her arms around his waist, and thrust against him a couple of times. It took awhile for the significance of this to sink in, but when it did, he turned around, grabbed the female by the waist, pulled her down so that she was sitting on his lap, and started to thrust. The leader of the group, who was lying only fifteen feet away, got up and approached them. The Outsider then retreated a few feet, and the leading male, Big Daddy, and the female sat down for a while together. Then he walked away, and the Outsider returned. Schaller says:

> She gazed into his eyes, and there must have been something in that look, for he did not tarry. With the female in his lap, he thrust rapidly, about twice each second, and soon emitted the peculiar call which I had heard during previous copulations. The female twisted sideways and squatted beside the male, who rolled onto his belly to rest in this position for ten minutes. Suddenly he sat down and pulled the female onto his lap again but she broke away. They then rested for a half an hour. After a long rest, the female got up again, stood once more by the back of the male, and he repeated the whole process again, and this time apparently bringing it to a climax.

Old male gorillas expect cooperation from their wives. Sir Julian Huxley has described an occasion when a game warden was watching through field glasses an old male ape, who was in turn staring at him through the undergrowth. The animal was too heavy to climb a tree to get a better look at the game warden, so he sent one of his wives up a tree to keep a lookout. "When her attention strayed, he reached up and pinched her bottom to remind her of her duty." The age of gorilla chivalry is not dead,

Yerkes mother gorilla, Paki, examines her just-born baby.
(Courtesy of Dr. Ronald Nadler.)

Paki nurses her newborn baby. The
umbilical cord and placenta are still
attached. (Courtesty of Dr. Ronald
Nadler.)

Female Yerkes gorilla presents sexually to male.
(Courtesy of Dr. Ronald Nadler.)

however, because there are stories of male gorillas lending a helping hand to females by holding their hand or arm when they are climbing an especially steep or rocky slope.

Unlike in the other apes, there appear to be no sexual jealousies among gorillas in the sense that an individual female is reserved for a particular male. There is a permanent leader of a gorilla group, usually an old silver back, but he does not seem to mind young males or even a male visitor outside the group having sexual intercourse with any of his females. Many human groups throughout the world have also been generous with their women to male visitors.

Although male and female gorillas mouth each other and may lick each other's genital regions, they do not engage in anything that can be called "kissing." In human beings the procedure of kissing has several connotations. There is a purely social kiss, most commonly on the cheek. There is also a lip-to-lip kiss, which is fairly common and is also given sometimes socially, although it usually signifies a sexual attraction between the two people exchanging it when they are endeavoring to demonstrate interest in each other. It is unlikely that any mouth-to-mouth kissing in humans is ever completely devoid of some sexual connotation. There is a sexual kiss with considerable oral exploration involving the insertion of the tongue into the opposite person's mouth and contact between the tongue and the interior of the mouth and the other person's tongue. There is, of course, the more complex oral exploration which humans are experts in, in relationship to the sucking, licking, nibbling, and light to hard biting of various intensities which some partners do to the genitalia of the other partner.

The tendency to use sex as a kind of a game is highly developed in man, but in the gorilla it seems to be largely concerned with the problem of reproduction. As the population pressure becomes greater in humans, and as it becomes more and more reprehensible to have a number of children and thus complicate life in the world, sex will become more and more a form of play activity unrelated to reproduction. There is no reason at all, of course, why this should not be so. There is a good deal of evidence in the last ten years that it is, in fact, becoming more and more of a national and an international sport.

It is interesting that presexual play in humans often involves the nibbling or biting of the earlobes. There are cases on record in both male and female humans in which orgasm has been reached simply from the stimulation of these appendages.

248

There is also a spongy type of tissue in the nostril which is erectile in the sense that it can accumulate blood and under certain circumstances becomes swollen. It certainly does when one has a cold or hay fever, but it also accumulates blood, becomes erectile, and makes breathing through the nostrils difficult during the process of sexual arousal so that even the nose has a sexual connotation as well as the earlobe. The interesting thing about the earlobes, however, is that we are the only primates that have them. No earlobes are present in gorillas or in any other great apes. It is difficult to understand precisely what their function is.

There is no evidence that the female gorilla enjoys any kind of an orgasm during sexual intercourse. This is probably so for the other types of apes as well. It may be that the development of the female orgasm in primates is related to the development of the erect posture. Apes tend to move around on their forelimbs rather than walk upright. Although they can and do walk upright on occasions, they usually do not do it for long periods of time. The same is true for monkeys. In animals who walk mostly on their forelimbs, the vagina is kept in a more or less horizontal position. Therefore, the female ape or monkey can become physically active immediately after intercourse because, having received the male ejaculation, the horizontally placed vagina will retain it. If the female, however, had immediately assumed the vertical position, there would be a tendency for the semen to drain at least to the lower part of the vagina, away from the womb, or else drain externally completely. If the human female resumed her normal position or normal activities immediately following copulation, the ejaculated semen in her vagina would probably drain out, too, since its opening is directed downward. This would greatly reduce the possibility of pregnancy. Desmond Morris suggests that here is an area where nature has intervened by providing an orgasm for the human female which leaves her relaxed and tired and sleepy and therefore likely to remain in a horizontal position for a considerable period of time after copulation. This is much more favorable to impregnation. If this is true, then the significance of the larger penis in the human male is understandable. The penis in the chimpanzee is long but very thin. In the gorilla, as we have pointed out, it is very short. Morris suggests that the human penis is large so that it will stimulate the vagina and the clitoris by stretching it a great deal more than any ape penis could possibly do, and in this way the stretching of the female

249

genitalia would produce the neural stimulation necessary for female orgasm.

Gorillas have a certain amount of sexual play. Some aspects of this have already been mentioned. Both sexes have been known to make oral-genital contacts, but they are not extensive, nor do they occupy much time. It is interesting that among humans the amount of sexual play and the amount of oral-genital sex and the amount of noncoital sex in general seem to be related to the social level and possibly to the degree of education and intelligence in the individuals concerned. Many human tribes have little presexual play in the way the more sophisticated know it, and in the uneducated section of the Western community there is relatively less sexual play, and coitus takes place quickly and with little preliminary activity. Thus the more uneducated humans are closer in their sexual reactions to those of apes. It is an interesting fact that many members of the community, particularly certain religious communities who regard sex as some kind of sin, think of sex simply as something to be used purely for procreation. The idea of enjoying sex with the other person and indulging in noncoital sex play is abhorrent to them. The individuals who framed the sex laws of many of the American states even forbade some types of presexual play. They forbade also sexual intercourse in other than the conventional position, with the male on top and face to face, which has been called the "missionary position." They have even specified that sex which occurs outside the bedroom is illegal. Apparently these individuals believed that in sexual relations we should behave like apes. The type of sex that can be obtained from most prostitutes, particularly those of a lower class, is the "in and out and off" kind typical of the apes.

We have mentioned the types of copulation between gorillas. Some of them copulate one behind the other in doglike fashion. In some cases there has been face-to-face copulation. Sometimes the male has lain on his back and pulled the female on top of him in a sitting position. Between humans copulation tends to be face to face, but is not exclusively so. There is a good deal of experimentation, including copulation from the rear. This is of special interest because there has been over the evolutionary period a considerable movement in the human female of the vaginal passage, which has swung into a more forward position. Nevertheless, in many human females it is still far enough back for penetration to take place easily from the rear. Man has been very inventive in developing positions for sexual

250

Yerkes gorillas in copulatory play. (Courtesy of **Dr. Ronald Nadler**.)

Mkubua copulates with Josephine. (Courtesy of Jim Ellis and Oklahoma City Zoo.)

intercourse (see the *Kama Sutra* and Alex Comfort), and again it is usually a more educated and intelligent type of person who is interested in this type of exploration. It is most often the simpler and less educated person or the one who is too busy who sticks to the biologically developed face-to-face form of sexual intercourse.

Masturbation occurs among the great apes, as it does in humans. Chimpanzees have been seen to do it with the hand or with the foot or even by rubbing the penis against inanimate objects. It is common in younger animals, but in an animal which is isolated it can also become much more frequent. Females masturbate with their companions. In them it seems to be a substitute for sexual intercourse. Masturbation usually disappears when mature, congenial animals who are sexually compatible have access to each other. Even baboons have been recorded as masturbating, and this is seen under both captive and wild conditions. It can be observed in young animals as well as in sexually mature animals. Again it is more common in males than in females. Male lowland gorillas have been seen masturbating in captivity. There was one gorilla known as Mambo, who was nine years old at the time he was observed and was housed with a mature female at the Bronx Zoo in New York. He was observed stimulating his penis with his hand, working it back and forth between the thumb and index finger. As he continued to masturbate, his eyes became rigid and had a kind of glassy appearance. There was also a lowland gorilla at the Milwaukee Zoo called Samson, who at the age of eight and in a cage with another male also stimulated himself with a hand for some minutes and thrust his pelvis several times. Neither of these animals was seen to ejaculate as a result of these masturbatory activities. Dr. John T. Emlen saw a female gorilla in the National Zoo in Washington playing with her clitoris. In the zoo in Basel a female gorilla has been seen to put her infant of three months between her legs and in a squatting position thrust at the baby with her pelvis at least ten times. Masturbation has been recorded in free-living gorillas.

The birth at the Yerkes Center of Kishina, whose mother was Paki, has been described earlier. Paki herself was taken from the wild when she was about a year old and came to the Yerkes Center and grew up here. She probably had little opportunity to observe and understand the role of the mother gorilla in bringing up the infant, so that when she had her first baby, Kishina, while she did not seem to understand very readily what she should do with it, she finally put it in a position where it was able to nurse and obtain milk. She nursed the baby and cared for it for about a week, keeping it almost continuously enclosed in her great arms. Then she seemed to become restless and dragged the baby around by one hand or by one foot, with Kishina complaining bitterly. Another female gorilla was put in the adjacent cage in the hope that this would make Paki feel more at home and less restless. She was then placed among other gorillas that she had been reared with and seemed to respond by showing a little motherly affection. A few days later Kishina was being mistreated again, and this time the baby was removed permanently from the mother. She was taken to the Yerkes great ape nursery to be reared.

Kishina is now growing up. She is about two and a half years old. Like young gorillas at that age, she is very prone to bite. This is probably part of their play behavior and is characteristic of them. On one occasion when the director was holding her up so that a photographer could take a photo of her foot, she put up with this for a little while, finally got fed up, and turned around and bit him on the chest. She did not break the skin, but she put some beautiful bruises and teeth marks on his chest. He went to his wife's office and bared his chest to her and said, "A gorilla bit me." She replied, "I must be about the only wife in the United States who would accept that story."

The removal of Kishina from her mother will handicap her, according to Dr. Ronald Nadler of the Yerkes Primate Center, in both her sexual and maternal behavior. She has had only a week with her mother, and she will be raised at the center and her behavior compared with that of other gorillas. Fortunately Paki herself has given us the opportunity to make such a comparison, since on April 26, 1974, she gave birth to Fanya, the father being Calabar, whereas Kishina's father was Ozoum. Paki seems to have learned from her experience with Kishina, and with Fanya she placed her in the right position for nursing very early and has treated her very well and with great care. Fanya has been about eight months with her mother now, and she will be a perfect comparison with Kishina.

Breast-feeding is important not only from a psychological point of view to the infant but from a nutritional point of view. It is known that the first material secreted by the mammary glands is important to the baby in helping it adjust its digestive system to outside food and that it possibly also has certain immunological properties that it transfers to the baby.

Maxine Rock, in an article in the Atlanta *Journal-Constitution,* recently compared her fifteen-month-old child, Michael, with Kishina, who was roughly the same age at the time of the comparison. The boy's weight was twenty-six pounds, a little smaller than Kishina's, and his movements were awkward and off balance, whereas Kishina's gait could almost be described as graceful. Kishina was able to walk when she was six months old, but Michael still progressed with a heavy crawl. She was not able to speak, however, whereas Michael had already managed "mama," "dada," and "bye-bye." The weight of gorillas at birth is surprising, three or four pounds, only about half the size of a human baby and yet capable of growing to a tremendous size, in the case of the male, at any rate, of 600 pounds. Paki's mother, when she produced the baby, was about 172 pounds. Maxine Rock quoted Robert Pollard, who is superintendent of the nursery at the Yerkes Center:

> But even if Kishina can't talk now, she does display a quick mind. She communicates with whines, whimpers, and grunts, and with gestures and facial expressions. She is tested every day by a researcher here; to me, the testing alone shows that this gorilla may be smarter than a two year old human. Why, Kishina can look at a picture of an apple, for instance, and match it to a real apple. And when she gets her meals, you better not give her two pieces of fruit when she is entitled to three, or one cup of spinach when she is supposed to get two cups. You can't fool Kishina; I believe she can count as well as I can.

Jimmy Roberts, the superintendent of the large-animal wing at Yerkes, said, "Even at her age, a toddler, Kishina has a certain dignity that you'd never see in another primate, including even a human child." It is true that even at the early age she was studied Kishina was more emotionally independent than a human baby, a point also noted by Maxine Rock. It is a gorilla characteristic that at the early ages they are more emotionally advanced than human babies. It doesn't mean that the animal doesn't need love and affection because it does. It early develops an inner strength probably related to the tribulations it meets with in the wild that a human baby doesn't pick up until it is three months older.

What is the day of a young gorilla like this like? Robert Pollard had the following to say about Kishina's day.

> We keep Kishina pretty busy; I don't think she has time to be

bored; we get her up at 8:30 for a morning check-up. At 9:00 it's time for her breakfast of pablum, milk and egg yolk or strained liver. As soon as she is finished eating, her cage is cleaned, and then at 11:30 she has a brunch of cabbage or carrots and whole wheat bread. She rests or plays until 1:00 when we take her outside. On a good day she stays out over an hour.

When Kishina goes outside to play, they take her chimpanzee cage mate. She is very attached to this cage mate and hates to be separated. The two have to be taken out into the play area outside together. Once outside, the chimpanzee is anxious to run about and to jump and play, but Kishina is much more settled. She seems to appreciate the environment around her more than the chimpanzee does, paying attention to the grass and to the wire and particularly to the people, whereas the chimpanzee is out exploring and playing and generally getting into mischief. Robert Pollard thought that in her behavior Kishina acted like a queen.

Dr. Adolph Schultz, distinguished Swiss-American primatologist, has a game with a little gorilla in the ape nursery at the Yerkes Center.

In New York's Central Park Zoo there are two gorillas, one called Congo, a male weighing 225 pounds, and a female called Lulu. On September 3, 1972, Lulu gave birth to a little baby, even though her keepers hadn't even realized she was pregnant. Her baby was called Patty Cake and made front-page news all over the world. Tens of thousands of children and adults came to see her. Lulu was delighted with her baby and for days refused to put it down. Incidentally, Lulu was eight years old and weighed 145 pounds. She rolled around on her back, cuddling the baby on her chest. She would kiss it on the head and sometimes kick her legs with pleasure. Congo, the father, was allowed into the cage with his wife and daughter, and both the mother and father were unusually tender and careful of the little Patty Cake. On March 20, 1973, Patty Cake put her arms through the bars in the direction of Congo, who was in the next-door cage, and Lulu pulled her back. As she did so, the baby's arm hit the bars, and it broke. She was operated on at the animal hospital in the Bronx Zoo and was able to receive twenty-four-hour care there. She was kept in a crib and changed regularly into Pampers and was given teething rings and toys to play with. In due course her arm mended, and the Central Park Zoo expected to get her back again. While she was at the Bronx Zoo, the baby had gained four and a half pounds in less than the two months that she was there. Her weight now was twelve and a half pounds. The Bronx Zoo felt it had done such a good job on her that she should stay there. The authorities at the Central Park Zoo, however, claimed that the gorilla was being spoiled by humans at the Bronx Zoo and that she should be returned to them forthwith. The controversy reached a deadlock, and about this time the authorities from the Bronx Zoo called the Yerkes Primate Center and explained the situation. It was suggested that they might ask the center's Dr. Ronald Nadler, who was working on gorillas, to come up and act as a mediator between the two groups. Dr. Nadler already knew about Patty Cake because on the night she was born he was telephoned by an enthusiastic friend in New York who knew that he was interested in gorillas and who called him up to inform him that New York had a new baby gorilla.

The controversy seemed in some ways frivolous. Dr. William G. Conway, the general director of the New York Zoological Society, said, however, that it was quite important since what they wanted was to do the best thing for Patty Cake because she had potential as a breeder. As he pointed out, the lowland gorilla is probably not going to come to the United States again because it is an endangered species. So if we are to help maintain the species and to continue to provide this species of animal for people of the United States to see, we cannot afford to waste a single animal.

After examining all the evidence and taking a week to think about it, Dr. Nadler finally decided that Patty Cake should go back to her mother and father. The reintroduction was made, and all seemed to go extremely well. Patty Cake was accepted by her mother again, and everybody was happy.

Humans appear to be obsessed with the possibilities of sexual relationships between human females and ape males. As we have seen, stories of gorillas running off with women are pretty widespread in Africa. In fact, wherever there is any kind of large ape, the stories exist: in Southeast Asia it is believed that male orangutans will pick up a human female and swing off through the trees.

When John Huston, the Hollywood actor and director, went on safari to Africa with a number of Hollywood female film stars to make a movie called *Roots of Heaven,* he said that not a single member of the group was ever molested by a gorilla or even solicited by a gorilla. Perhaps the gorillas were not impressed by what they saw; after all, a big gorilla, even a male, has a chest measurement of seventy-six inches, whereas the best of the glamour girls does not exceed forty inches by much. Rocky Marciano, the boxer, as a matter of interest, has a chest measurement of forty-eight inches.

At the Yerkes Center we have seen both male and female apes develop an attachment for persons belonging to the staff at the center, not necessarily, however, with someone of the opposite sex. Some human males can cause a male chimpanzee to show an erection just as often and just as quickly as a female human can, and the production of an erection in the chimpanzee often seems to be more associated with excitement and seeing a new face, male or female, than with the presence of someone of the opposite sex. Gorillas do not produce erections so readily.

At one of the Swiss zoos a young female employee was leaving late one evening after everyone had gone home and noticed that one of the gorillas appeared to have caught its hand in the wire of the cage. Unlocking the cage, she came in to see what the problem was. The door, which was self-locking, slammed with her keys in the outside. She found herself locked in the cage with the gorilla for the night. The animal, however, who was in no difficulty, was very happy to have her company and placed his arms around her and took her over to the side of the cage, where he lay down and slept with his arms around her all night. He made no attempt at any sexual play of any sort. After the gorilla fell asleep, the girl thought this was the time to loosen the giant hands and withdraw from his hug. Every time she tried to do this, however, the gorilla stirred and clasped her even more firmly to him. In the morning they were still cuddled together when the members of the zoo staff arrived. They were amazed to see what had happened. The girl was disheveled, of course; her clothes were torn, and she was in a state of shock. Otherwise she was completely unharmed. It seemed that the gorilla was just lonely, and, in fact, our experiences with gorillas indicate that they often have this longing for companionship.

Not so long ago one of the Yerkes animal caretakers was locked in a cage with one of the female gorillas at lunchtime when there was no one around. He kept calling for help, but no one heard him, and he was compelled to spend the lunch hour in the cage with the animal. This was a big female gorilla which had had some association with humans earlier when she had been in the employ of the organization known as Africa, U.S.A., which is located in California and which provides apes and other animals for films for television. Her name is Shamba, and she is now a big girl of probably 170 pounds or so and quite mature sexually. She showed absolutely no interest at all in the member of our staff who was locked in with her and, in fact, for the whole of the hour she behaved as if he were not even in the cage despite the noise he was making trying to get out.

The desire for companionship by gorillas is also shown by an occasion when the director of the Yerkes Primate Center went into a cage with two Yerkes gorillas, Paki and Oko, two females who were partly grown youngsters. As soon as he entered the cage, they both came over to be stroked, cuddled, and embraced. He spent probably ten to fifteen minutes with them and gave them the little bit of food he had. When he got up to go, he found that one of them had folded her arms around his waist and the other had encircled his legs with her arms. Between them they had inactivated him, and it was almost impossible for him to move. He hobbled, with the two of them hanging onto him, over to the door and finally managed to extricate himself and get out. Even so, he probably could not have done this had not someone been waiting outside who was prepared to unlock the door for him at the right moment and let him slip out. If these animals had been much bigger, he probably would not have been able even to shuffle over to the door of the cage.

To continue with the story of a relationship between apes and humans, there are reports that Trader Horn, who traveled for many years in Africa, knew of a case in which a white man caged a slave girl with a gorilla, having heard that an ape is very attracted to human females; this man expected that the gorilla would rape her, but the animal made no physical contact at all. He remained in one corner, and the girl cried in the other, so there was no hoped-for result. It was rumored that the man who did this was later shot by other whites, who were horrified by the story of what he had done.

During the Middle Ages and even later there were reports of very hairy humans who were sometimes thought to be the product of sexual relations between a monkey or an ape and a human. Edward Topsell talks of what he described as "satyr-apes," which he said were shaped like humans, had a lot of hair, lived in a sultry condition, and had a great desire for sexual intercourse with human women. These satyr-apes seemed to have had their origin in the stories of various sailors. Mythological satyrs resemble goats more than apes. They had the goat's rear legs and hooves, and their bodies were covered with hair. It seems likely that some of the earlier observations, especially of the apes and possibly also baboons, resulted in their being interpreted as some kind of satyr. Baboons are also supposed to be licentious and are said to attack both women and children with the object of copulating with them. Sir Richard Burton, in one of the stories he translated from the *Arabian Nights,* describes an Abyssinian baboon which is said to have attacked a woman in a Cairo street and tried to rape her. It was then killed by being bayoneted by a sentry. If such an incident did occur, it was probably a straight-out attack by the animal on the woman which was interpreted as a sexual attack. In the Roman arena baboons are also reputed to have been used to rape little girls. In another of the tales Scheherazade was reputed to have taken a baboon as a lover, which spent the day with her doing nothing but eating, drinking, and copulating.

The Beauty and the Beast story is also found in the *Arabian Nights.* One of the stories that Scheherazade tells is about a butcher and the daughter of the vizier. Warden was a royal butcher who was secretly in love with Ladite, who was the young daughter of the vizier and only fifteen years of age. Every time she went out he would follow her, and eventually he found that each night she went to the basement of the palace. He was both curious and jealous because he assumed that the young lady had a lover, and he was determined to find out who he was. Equipped with an enormous butcher's knife, he checked one night on his suspicions. As he peered through the crack in the door, he was astonished to discover Ladite engaged in a variety of different love games with an enormous gorilla. The love games, which we will not describe here, are actually described in great detail in Scheherazade's narration. Warden, however, being very upset by this sight and during a pause in their love activities, threw himself at the gorilla and cut its throat. It is assumed that because the daughter of the vizier was so precocious, the gorilla was, by this time, fairly exhausted and not able to defend himself very well. Ladite reproached Warden for killing her gorilla but then told him the history of her relationship. Ever since she had been very young, she had been under the care of a gigantic black slave. He had initiated her to sexual activities, but he had a fatal accident, and she was advised that the only substitute for a black giant was a gorilla. She purchased a large gorilla from a merchant which had shown a peculiar preference for her when she encountered it in the market. Warden convinced Ladite that whatever an ape can do, man can do better, and he persuaded her to marry him. Scheherazade does not tell how the butcher stood up to the eroticism of the beautiful Ladite, but Francisco Montaner, who wrote this story, says that it was his assumption that Warden did not last very long with such a highly sexed young lady.

The human-ape sexual relationship was also the theme of the famous stories of Tarzan by Edgar Rice Burroughs. Tarzan, the son of Lord Greystoke, is stolen in infancy from his father and mother in the African jungle and grows up with the chimpanzees who captured him. Later when he sees an American girl, Jane Porter, he immediately falls in love with her. After he has seen her, she is attacked by an ape, who apparently plans to take her for himself. The ape throws her across his shoulders and leaps into the trees with her and attempts to make off, presumably with the object of copulating with her and making her his permanent mate. They are, of course, chased by Tarzan, who catches them and kills the marauding ape and so wins the love of Jane.

260

In view of all the stories about apes and human women, one might ask, have there ever been any hybrids between humans and apes, or have there ever been any hybrids between the apes themselves? To this we cannot give a definite answer. There is no real evidence that a successful cross has ever been made between a human and an ape. But there is no reason at all from a chromosomal point of view why, for instance, a gorilla mated with a chimpanzee should not produce living babies. And it seems reasonable to suppose, in view of the evidence we have had from hybridization of other animals, that even the orangutan might hybridize with the gorilla and the chimpanzee. The orang, however, has a method of sexual intercourse rather different from that of either of the two other apes. Under natural circumstances it has at least some sexual intercourse in the trees, even while hanging from the branches. So mechanically it

Distinguished primate anatomist Dr. Helmut Hofer of the Delta Primate Center plays with baby gorillas in the Yerkes nursery.

might be difficult to secure copulation between the orang and either the chimpanzee or the gorilla. But in the state of captivity the new techniques of artificial insemination might make it possible for such hybridization to be carried out without too much difficulty. There is a rumor that there was a cross between a chimpanzee and a gorilla in Georgia many years ago, but we have not yet succeeded in tracking it down to its source. As far as humans and apes are concerned, there seems to be very little physiological reason why artificial insemination could not be used between humans and apes with the possibility that a viable child might be reproduced. Gorilla and chimpanzee chromosomes are like human chromosomes and similar enough in number to those of man to make such a hybrid conceivable (in both senses). It is surprising that this type of hybridization has not in fact already taken place. What such an animal would be like and what the moral significance of this would be, it is very hard to say. No doubt it would provide some problems for the ecclesiastics in deciding whether an animal such as this would have a soul or not.

The theme of the gorilla having sexual feelings about a human or relations with a human female has been carried through into art, literature, and moving pictures. That and the horror theme are the two most characteristic themes in movies that have to deal with gorillas.

13

"Some of My Best Friends Are Gorillas"

THE Yerkes Primate Center now has fifteen gorillas, eleven females and four males. Ozoum is a big beautiful male lowland gorilla which was purchased in 1961. At that time he was a sweet little baby that used to enjoy cuddling, but now he is too big and rough for that tender sport. In 1961 we acquired seven more gorillas in our family: Calabar and Rann, who are males, and Choomba, Jini, Oko, Paki, and Segou, who are females. Then in 1964 Banga and Oban, a female and a male, arrived. Inaki came from the National Zoo in Washington, D.C., in 1967, and in 1969 we got a movie star; her name was Shamba; she had appeared with Clarence, the cross-eyed lion, in a film called *Daktari*. Shamba was obtained from a farm for movie animals called Africa, U S.A. In 1972 Yerkes had its first gorilla birth with the happy arrival of Kishina, a female. In 1974 another female was born at Yerkes. Her name is Fanya.

In case you've wondered, at the Yerkes Center all female apes are given names which end in a vowel, and all males are given names which end in a consonant. With the exception of the center's first gorilla, all gorilla names are Swahili words.

The Yerkes Center's first gorilla was Mary Lou, who was brought to the center around 1959, when Dr. Arthur Riopelle was director. She was paralyzed down one side of her body. Mary Lou had been in a hurricane in Florida when she was a tiny baby and had been hit on the head by part of a building when it collapsed. This had left permanent damage to her brain. After a few years she began to develop epileptic seizures, and we brought in one of the best neurosurgeons in Atlanta to examine her. He decided to open up her skull and see if there was any obvious defect in the brain which could be remedied that would help her get rid of her seizures. Unfortunately, when he got down to the brain, it was obvious that one side of it was seriously degenerated, and nothing could be done. Mary Lou lived on for a few more years and finally died—much to our sorrow. She was a lovely animal, and it was fun to go into her cage and pet her. She used to enjoy it so much, but after a time used to get worked up emotionally, and it became obvious that was the time to leave because it was impossible to tell what would happen next since she was still quite strong despite paralysis.

Ozoum is now one of the biggest gorillas in the center; he was sent by an American animal-import company. He arrived about half past two in the morning by plane and weighed about twenty-five pounds. He was about one year old then. The other big

Bandam, young Yerkes gorilla that developed polio, moves around nursery in child's walker.

Bandam receives hydrotherapy in the Yerkes nursery.

gorilla is Calabar, who was born in Africa toward the end of
1963. He arrived at the Yerkes Center early in 1965, and he
weighed about thirty-two or thirty-three pounds. He came
from the Far East Animal Imports Company, which is located
in Tilburg, Holland. He arrived with a full set of baby teeth.
His companion on his journey from Holland was Rann, who
was probably born about the same time. We do not, of course,
know the birth dates exactly since they were born in the wild.
Rann was about twenty-five or twenty-six pounds when he arrived. Within a year Calabar had gone up to forty-two pounds.
Calabar has now outstripped Rann considerably in total weight.
Though Rann's rate of growth had been impressive and in
another four years he weighed over 100 pounds, he is still not
as big as Ozoum or Calabar, and he is a rather less cantanker-
ous and more gentle animal than those two.

Another male arrived on October 28, 1964, landing at the Jacksonville Airport at 8:45 P.M. He had been fed in Amsterdam at 8:00 P.M. (Holland time). He had been given a bottle of milk and some biscuits just before he departed from the airport. He weighed only ten pounds on arrival, so clearly he was very young. When he arrived at the laboratories from the airport, his pulse was 180, and he was very tired and was fed with the bottle again and given a little bit of fruit and some water. He was left overnight in the box in which he came and in which he seemed to be very comfortable.

Of the two big gorillas, Ozoum and Calabar, Ozoum is probably the more excitable and rough. While the director was talking with him recently and scratching him through the bars, the animal turned and swiped the man's right index finger, almost breaking it off at one of the joints. It has now, fortunately, healed.

Katoomba is a female gorilla that had the distinction of being photographed while cuddling with Dr. Boris Lapin, the director of the Russian Primate Center, when he visited the Yerkes Center. Katoomba is estimated to have been born around July, 1962, and she arrived late in 1963, weighing twenty-five to twenty-six pounds. The superintendent of the center, Graham Townsend, and his wife, Murray, drove to Orlando one night to pick Katoomba up. She had flown from Amsterdam on an SAS jet and then came down from New York on Riddle Airlines. She was brought to the laboratories late in the evening and given some fruit and cooked sweet potato. She ate very happily, but she refused a cup of milk. Katoomba was a nice little animal to begin with, but she too rapidly learned to bite and became a little bit of a problem. Paki and Oko arrived together on July 17, 1964, in company with two other babies. Both Paki and Oko are now a fine pair of female gorillas.

Another animal which is good company is Jini. She was estimated to have been born around November, 1963. She also came to us via the Far East Animal Imports Company, on November 20, 1964. She had been given some fruit and a drink of sweet tea, of which she was very fond, at 7:00 A.M., immediately before the departure of the plane from Holland to New York. She was a sweet little animal, weighing twelve to fifteen pounds, and arrived at 4:30 P.M. at the Jacksonville Airport from New York after the flight from Holland. Apparently there had been a great deal of trouble and delay in New York in securing her transfer from the jet in which she had crossed the Atlantic to

Portrait of Mary Lou, Yerkes gorilla that received a head injury during a hurricane in Florida when she was a baby and was permanently paralyzed down one side of her body.

267

Yerkes gorillas enjoying their new compound. (Photo by G. Bourne.)

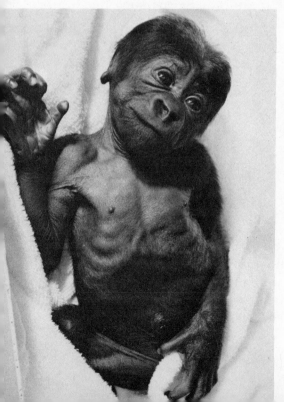

Yerkes firstborn gorilla, Kishina, two days old. (Photo by Yerkes staff photographer Frank Kiernan.)

Yerkes gorilla training as backstop for Atlanta Braves. (Photo by Yerkes staff photographer Frank Kiernan.)

the plane which was to bring her down to Jacksonville. She arrived at the laboratories at about 5:15 P.M., very hungry, and was fed a bottle containing milk formula and given some apple and orange, and she settled down to sleep very well. A few years after her arrival she was given some tests by Dr. Bernard Rensch and Dr. G. Duchen from the Zoological Institute of the University of Münster in Germany. They found that she learned quickly to move books, untie loops of threads, wind up a chain, open the door of a wooden box, remove a metal stick from a hole, and remove covers from little boxes of Plexiglas or aluminum.

We also had a gorilla called Bandam, who developed polio. Bandam lived for some years in our ape nursery. He used to move around in a stroller. He actually had one arm almost completely paralyzed and both legs partly paralyzed. He had been purchased some years ago while our laboratories were still in Orange Park, Florida. A few days after he arrived from overseas, he developed a high temperature and shortly after that showed signs of paralysis. We were highly suspicious of the circumstances and decided that this might be poliomyelitis. We got in contact with the National Center for Disease Control in

269

Atlanta, and a team of experts in this disease came to Orange Park within twelve hours to survey the situation for us. By this time it was evident that another gorilla had the disease, and so did one of the orangutans. As soon as the disease was completely diagnosed, all the other animals in the colony were hastily given polio vaccine. This was in 1964. It was a small epidemic of poliomyelitis, and it is of interest that the only epidemic of that disease that occurred in the United States in that year was the one that broke out in the Yerkes Laboratories. This, incidentally, was the first time it had been shown that this disease occurs spontaneously in great apes and that it could be transmitted from one to the other. One of the gorillas that got infected with poliomyelitis died; Bandam survived, but in a partly paralyzed condition. He was kept continuously clad in diapers. Later when we came up to Atlanta from Orange Park, he was put in one of those little walkers in which young babies are placed in an attempt to teach them to walk. He could move around the nursery in this walker, which had four small, well-oiled, easily running wheels, by moving his feet against the floor. He was, of course, a great favorite with everybody and gave every indication of enjoying the love and attention that was lavished on him. He was usually put in the walker in the early morning and allowed to run around the nursery all day as he liked. He followed the nursery workers from room to room and was given special food. Unfortunately, or perhaps fortunately, Bandam did not survive more than a few years; he would have been a terrible burden to himself and an impossible one for us to handle if he had grown to adult weight.

On February 28, 1968, we received Inaki, who was born at the National Zoological Park in Washington on April 8, 1967. She was a little under one year old when she came to us. She had been raised in the home of her keeper, Bernie Gallagher, and was actually looked after by Mrs. Gallagher. At birth she weighed about four and a half pounds. We exchanged Seriba, our firstborn orangutan at the center, for her. The Yerkes Center has been extraordinarily successful in raising orangutans, having produced over thirty of them at the time of writing. Gallagher brought Inaki down, sitting on his lap in a first-class seat, in an Eastern Airlines jet. The other first-class passengers completely ignored the animal, probably feeling annoyed that it was allowed to be in the first-class section. The man sitting alongside Gallagher did not address a single word to him about Inaki, who was, after all, less trouble than a baby because at least gorillas don't indulge in the fits of crying that human babies seem inevitably to do when they are on an airplane. A man who was reading the paper in the seat across the aisle looked up and saw the animal and asked what it was. Gallagher told him it was a gorilla. He said, "Really," and went back to reading his paper, and that was the only attention Inaki got on the flight down except, of course, from the stewardesses, who fussed over her and petted her all the way down to Atlanta.

The reason why the animals had been coming in the Jacksonville Airport or the Orlando Airport was that at the time we bought most of our gorillas, the laboratories were still down at Orange Park, Florida, about twenty to twenty-five miles south of Jacksonville. By the time we bought Inaki we had already been established in Atlanta, Georgia, on the Emory University Campus for some three years.

Inaki began cutting two teeth at twenty-seven days. She adapted extremely well to the center, having been put in the nursery and having been given all the love and care and attention that all our baby apes get. Later on she was used by Dr. Russell Tuttle for his studies on the action of muscles. Gorillas, like chimpanzees, walk quadrupedally, as we pointed out, and they close their hands and put their weight on the knuckles of the front limbs. Dr. Tuttle was particularly interested in the muscles which were used in this procedure, and he trained Inaki to receive some tiny electrodes into the muscles, which, of course, caused her no pain. These he connected to a machine which recorded the electrical action of the muscles while Inaki was walking. Inaki enjoyed these sessions and collaborated very well with Dr. Tuttle and Robert Pollard, the nursery superintendent, playing with them and tumbling over and over and paying very little attention to the lead that went from her arm to the machine in an adjacent room which was recording all the things that her muscles were doing. She proved a very cooperative animal, and we are sure she enjoyed these sessions very much indeed.

The two big events among our gorillas, of course, were the births of the first two gorilla babies at our center. Paki was the mother of both the new babies. The first baby, Kishina, a name which comes from Swahili and means "the source" or "a kind of dance," was born on August 7, 1972. Paki's labor and birth have been described earlier. The father of Kishina was Ozoum, and the length of gestation was between 249 and 250 days. She la-

271

bored only two and a half hours. The birth of Kishina was recorded on videotape. This is the first time in history that the birth of a gorilla has been recorded in this way. Paki accepted the infant very well and cleaned it and placed it on her stomach and chest. The infant was in good condition, was vocalizing, and was able to support its own weight with its arms. The placenta remained attached until the following morning, when the cord was presumably bitten through. A couple of days after the birth it was necessary to anesthetize Paki because some of the membranes had been retained in the uterus and had to be removed by the veterinary surgeons. The baby was eventually taken from her mother, who started out being a very good mother, then suddenly began abusing her baby, throwing her across the cage floor and rolling and tumbling her as though she were an object or toy. Had we not taken Kishina away, she would not have survived.

On April 26, 1974, Paki had another baby, called Fanya, a Swahili word which means "to give" or "to produce." In the case of Fanya, the period of gestation was 245 to 247 days. The father this time was Calabar. The birth was normal, with no complications. In fact, Paki could not even be seen in labor, and she produced the baby this time in a squatting position. Paki began nursing the baby within twenty-four hours, and she was certainly a great deal more proficient with maternal activities and nursing behavior than with the first baby. With Kishina she had spent at least an hour after birth positioning and repositioning her infant, and this time she placed Fanya on her chest immediately after delivery, and it did not take Fanya very long to find the teat and begin to suck. She was nursing well within twenty-four hours.

It is ten years since the Yerkes Center bought any gorillas. For one thing, they are on the endangered list, and it is illegal. We also would not want to take any more of these animals out of the wild. We feel we have a viable breeding group now, and we can continue the species with the animals that we already have. In the nursery the baby gorillas, until they are able to sit up, are kept in incubators, clad in diapers, and are fed a routine bottle with formula and later on are given food supplements. As they get older and are able to sit up and totter around, they are put in small cages in one room. Usually we try to put them with a companion. Since we don't have many young gorillas, it is often a young chimpanzee or an orangutan that they have as a companion—one about their own age. They adapt extremely readily to the new companion and get on well. There is no racial discrimination among the baby apes.

As they get a little older, the babies are put in another room, where the cages are bigger, and they do not go into this room until they are strong enough to stand up and climb around. The animals in these rooms are taken outside once a day when the weather is good and are given an hour or two to play on jungle gyms situated in a couple of grass-covered exercise yards. They go out in groups, but since we do not have groups of young gorillas yet, Kishina, for instance, would share an exercise yard with a number of orangutans and probably some chimpanzees as well. When they have outgrown the nursery stage, the animals go out into the large-animal wing, where we have a number of large cages divided into two parts, an outside section and an inside section. The outside section is about ten feet square and is open partly to the elements, but the overhanging roof gives some shade if the gorillas want to be outside but still sit in the shade. If they want to sit in the sun, they are able to do this also. The inside run is separated from the outside by a guillotine door which the animal can raise or lower or which we can lock in place if we want to keep it inside when the weather is too inclement or too cold. The inside quarters are heated, but not air conditioned, since this is not necessary; even on the hottest Georgia day the temperature rarely gets above 75° in the large-animal run. It sometimes gets very cold in the winter for short periods, and we have these inside areas heated all through the winter. There is a slab mounted a couple of feet off the ground which the animal can use as a sleeping perch. Most of them do, in fact, sleep on this perch. The inside area is about eight by ten. Sometimes we have two animals together in one of these little suites, sometimes only one. A couple of cages at one end of the run have now been joined together so as to make a much larger suite where we can put some of the bigger gorillas as this gives them much more room to run around and entertain themselves.

Out at our field station, where we have 110 acres of countryside, we were fortunate to have a small stream running through our property. At the field station we have a number of compounds for monkeys and apes to run around in and to enjoy themselves and live a social existence . The small stream running through the property was dammed up some distance downstream by the authorities, and this provided a twelve-to-fourteen-acre lake on our property. Before the dam was closed, we arranged for a contractor to scrape up the dirt with a bull-dozer and make two islands. On these islands we have placed a group of chimpanzees and another group of orangutans. We have a well-constructed compound in which we are placing a number of gorillas. Five female gorillas have already been sent out, and we plan to add two or three males to this group at a later date. The females were released in the autumn and thoroughly enjoyed the freedom of the 100-square-foot compound with three sleeping compartments. They all ran happily around the compound; in parts where there was some mud they slid on it, slapping one another on the back as they ran past with a slap that resounded all over the 110 acres of the field station property. They clipped one another over the head as they went past, pushing one another and generally having a wonderful time—all but Inaki, who, having been brought up as a baby in a house with humans, still felt herself more closely identified with humans than with gorillas. Inaki crouched in a corner and resented the approach of any of the other gorillas. She later withdrew altogether, staying inside one of the sleeping quarters and refusing to come out. We finally found it necessary to withdraw her and sent her back to her home at the center.

After the female gorillas had got used to the compound, they were removed and replaced by three males, who were equally happy at being able to run around and slap one another on the back. The males included a big tough silver back known as Calabar, who immediately asserted his superiority. As soon as the males felt at home in the compound, four female gorillas were released into it. There was a lot of wild excitement, with Calabar pounding his chest and running at the females. The latter were finallly intimidated into seeking refuge inside a climbing facility made of metal pipes. There they were at least safe from the runs and heavy slapping that the males, especially Calabar, handed out to them. After a few days of this, however, the females rebelled and three of them caught Calabar in a corner and taught him what women's lib is all about. Poor Calabar had to be removed from the compound with a deep slash in his chest, a cut on his ankle, and another deep cut across his leg; he is now recovering from his wounds at the Yerkes Center. One of the other males also got sick (we can't say if it was psychosomatic or not) and had to be removed from the compound. The remaining male, Rann, when he was young, had spent a lot of time being caged with female gorillas, so he presumably knew how to handle them or knew how to let them handle him.

At the time of writing he is living more or less peacefully in the compound with his four female friends.

We anticipate that a group of animals with this additional freedom will be much more productive in producing baby gorillas. Until recently it was the biggest captive group of gorillas anywhere, but now the zoo at Cincinnati, as we indicated earlier, also has a group of gorillas of this size.

Although there are some hundreds of gorillas in captivity throughout the world, they are located almost exclusively in zoos, and very little scientific research is carried out on them since few zoos have the facilities or the staff for such activities. The Yerkes Center has, however, one of the largest groups of gorillas in captivity, and it is available for various studies. It must be remembered, however, that gorillas are on the endangered list, and there is no excuse for carrying out any kind of experimental procedure which is likely to harm the animal or endanger its life or even affect its breeding capacity. In fact, this philosophy has to be applied to experimentation with any of the great apes.

There have been a number of studies of gorillas in the wild, in which their day-to-day behavior has been recorded, the social relations among the individuals in a group have been listed, and their diets and vocalizations have been studied. This type of investigation has not, however, been carried out on a group of captive gorillas, of which closer and more intensive study can be made. Our gorilla group at the field station will provide us with an opportunity to do this.

Sexual behavior of gorillas has been observed in the wild, but there have not been many observations recorded. As mentioned earlier, Schaller saw only two copulations during the whole of the time he observed these animals. Planned mating of pairs of gorillas, particularly when a female subject is at the peak of receptivity, has permitted Dr. Ronald Nadler of the Yerkes Center to carry out more than 2,000 thirty-minute mating tests providing the clinical observation of gorilla sexual behavior as an aid to learning new techniques of breeding. Studies of maternal behavior have led Dr. Nadler to make the following observation:

> Preliminary assessment of these data suggest that the early rearing of this gorilla mother with a group of peers better prepared her behaviorally for her first maternal experience than would have been the case had she been reared alone. More-

275

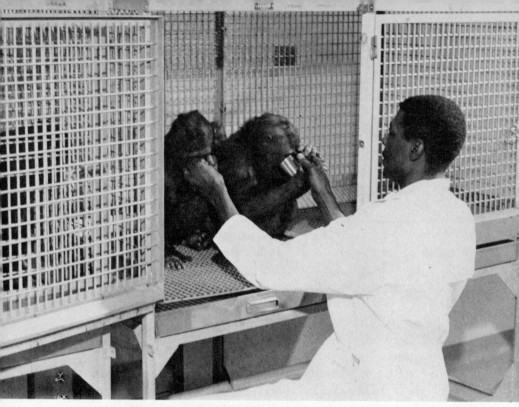

Young Yerkes gorillas get their morning drink of milk. (Photo by Yerkes staff photographer Frank Kiernan.)

over, it appears that her first delivery provided valuable specific experience that facilitated development of the more adequate maternal care she demonstrated with the second infant. This information, together with a few other reports of a similar nature on gorillas in zoos, is of special interest because of the endangered status of gorillas in the wild. It suggests that small, self-perpetuating colonies of gorillas may be maintained in captivity for prolonged periods of time and be available for replenishing wild populations if eventually that becomes necessary.

We had expected our gorillas to breed more readily than they have so far, but as Dr. Nadler points out, all our gorillas were separated from their mothers and came into captivity at an early age. Studies by a number of investigators have shown that primates need close contact with the mother and with their own peers during the early years of their lives to be able to have proper social interaction with other members of their own species. It may be this isolation which has made our rate of gorilla reproduction so slow, but the same criticism may be made of our

276

orangutans, which all came to us as youngsters, but they have bred extraordinarily well.

Further studies of a sociological nature are now under way. Dr. Nadler has initiated some studies with a graduate student, Robert B. Fischer, of the University of Georgia on the establishment of dominance by gorillas. Monkey groups have a very strong dominance hierarchy, in which every monkey knows his place and dares not try to change it. Hierarchies are much looser in structure in the apes, but we know least about them in the gorilla.

The Yerkes gorilla collection also gives us the opportunity to study the hormonal changes which take place in the sexual cycle of the females; such information has already been obtained in detail for the chimpanzee (they are very close to humans in this respect) and to a lesser extent for the orangutan. It would be of special interest if the gorilla hormones were identical with those of humans because of the anatomical similarities between the two.

Gorillas, it has been observed, seem to be especially susceptible to arthritis. The National Zoo in Washington has an eight-year-old gorilla, Tomoka, who had a serious case of arthritis, which was eventually cured by doctors using an antibiotic. The condition had been diagnosed as rheumatoid arthritis.

A Washington doctor, Dr. Homer Brown, has believed for a long time that rheumatoid arthritis is caused by a germ called mycoplasma. These organisms are very small in size and difficult to isolate in man. Dr. Brown, in view of the fact that he has considered arthritis to be infectious, has been treating the condition in humans with antibiotics. There was no way in which he was able to tell why the treatment seemed to work, although he had had a good deal of success with it. The gorilla Tomoka was born in the National Zoo in 1961. He appeared to be a normal animal, very playful, until he was five years of age. Then he showed signs of a problem in his joints and muscles. His growth rate and his body weight both stopped at this point, and his personality seemed to change to one of depression. When the doctors first saw him, the arthritis had attacked his left foot, hand, and wrist. A test of his blood indicated that there was a positive rheumatoid factor there. Mycoplasma was also isolated in his throat and also found in the tissue in his infected joints. This was a new strain of mycoplasma and was named the TK strain, after Tomoka. It induced the production of antibodies when it was injected into rabbits used for this purpose. Zoo officials had

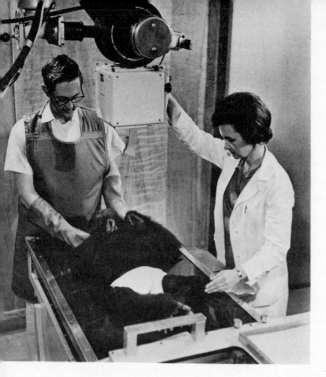

Young Yerkes gorilla receives chest X ray to help the veterinarian diagnose tuberculosis if the animal should be accidentally infected.

Gorilla babies in Yerkes nursery.

used twenty-nine different medications to treat the disease, but these had had no effect. Dr. Brown and his team turned to the antibiotic tetracycline. This attacked the mycoplasma, and the condition was dissipated. Two and a half years after this event Tomoka seemed to be in good health and appeared to have returned to his previous good spirits.

The Yerkes Center has already made studies of the blood cells and the chemical composition of the blood of gorillas and has found them practically indistinguishable from those of humans. Gorillas also have blood groups like humans, and immunological studies continue to be made on them. Such studies involve the gorilla in no pain or danger and they can be done during a time, every six months, when the animals are anesthetized for a chest X ray to be sure they have not developed tuberculosis. This disease is one that is greatly feared by all who keep gorillas in captivity and was the principal cause of death of the gorillas which have been brought to the West from the earliest times. We take the most elaborate precautions to ensure that our animals do not catch the disease, and if they do so, we intercept it at the earliest possible stage. To implement this, we make every new employee and every scientist who wants to work at the center produce a certificate from a doctor saying that his chest X ray and skin test are negative for tuberculosis. Then every six months every member of the staff, whether he or she has an association with the animals or not, has to have a skin test and a chest X ray, and so do the apes themselves. As a result of these precautions we have to limit the visitors we accept at the

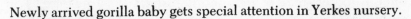

Newly arrived gorilla baby gets special attention in Yerkes nursery.

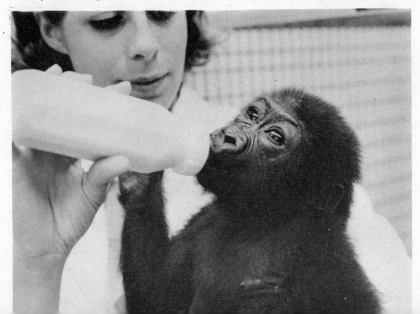

center. The general public cannot be admitted at all, and other visitors are not permitted to get close enough to the animals to infect them. Zoos are also worried about their apes catching tuberculosis, and gorillas, chimpanzees, and orangutans in zoos are usually seen either behind glass or separated a considerable distance from the public by moats.

Among our other research activities with gorillas, we studied the early physical and mental development of the gorilla babies, and Dr. Donald Robbins of the Emory University Psychology Department studies memory in gorillas. They have also been used in a variety of behavioral and mental tests by Dr. Duane Rumbaugh, who finds the gorilla's intelligence on a par with the chimpanzee and orangutan.

Our veterinary staff watches over our gorilla group with hawklike attention; the slightest sign of malaise starts everybody working, from the veterinarian to the clinical laboratory technician. We do not hesitate to bring the most distinguished physicians to consult with us if we feel our gorillas need it; nothing is going to happen to the Yerkes gorillas if anyone can prevent it.

We hope that they will form a self-sustaining group that will perpetuate this noble animal for all time.

This may be an appropriate spot to record a hilarious story about the Yerkes gorillas. Recently a wild-animal park called Lion Country Safari opened up south of Atlanta. A lake with three islands was constructed on the grounds. Since we had been rather pressed for accommodations for our apes, we offered to put some of them on the islands—a pair of apes on each one. One day three workmen on a raft paddled over to the gorilla island to carry out some minor maintenance; one of them went ashore, and the other two stayed on the raft a few feet away. One of the gorillas, Rann, saw the workman on the island and charged him; the workman looked up, saw the animal bearing down on him, and promptly dived into the lake. By the time Rann got to the edge of the lake he was going so fast he could not stop and went on into the water too. He managed, however, to flounder to the raft, and as he grabbed it, the two men on it dived into the water. Rann climbed onto the raft and gravely surveyed the humans milling around in the water. Eventually the humans pushed the raft back against the island. Rann stepped ashore, the workmen climbed back on the raft, and all was well.

14
Monsters—a Wild and Hairy New Breed

WHEN gorillas were first discovered by Western-
ers, some 130 years ago, mankind was satisfied to know that
there was some hideous monster tucked away somewhere in
darkest Africa about which it could fantasize. Perhaps this
reflects some kind of racial memory, a link with the past. When
modern humans, at least Cro-Magnon man, developed, in the
areas where they lived there were almost certainly groups of
Neanderthal men who must have seemed to them like monsters; it is quite possible the fear of these monsterlike creatures
in the woods is an inherited racial memory.

Man's belief in monsters or ogres dates back to early historical times. In medieval Europe the devil was the principal ogre,
and when the gorillas came along in the late 1800's, it simply
provided an alternative to the devil. As we have observed, the
arts and motion pictures, always in search of what seems to satisfy the desire for people to be frightened, selected the gorilla
as a monster who ranked, in the chamber of horrors, above the
werewolf and the vampire bat. Another primitive theory insists
that the big, hairy, manlike beast carries off a human female
and enjoys some kind of sexual relationship with her. Man,
fickle in nature, discards monsters when they no longer fright-
en or fascinate him. The gorilla has tended to fade now as the
ranking monster. It is rather common in zoos, and scientific
knowledge has somewhat deflated the monster myth, so man
seems to have needed to look for another type of monster. The
Abominable Snowman and Bigfoot were perfect substitutes.

In the Eastern part of the world, in the area of the Himalayas, the search for a local monster, a genuine homegrown
product, has existed for more than 100 years.

In fact, there had been a legend in the Himalayan Mountains for centuries that strange creatures covered with hair and with a humanlike appearance had been seen and had left tracks in the snow. In 1887 Major Lawrence Waddell, who was a major in the medical corps of the British Indian army, wrote a book (published in 1889) in which he described many of his explorations in the area around Sikkim, a small state between India and Tibet. He was impressed by the fact that the tracks that he saw in the snow had been made by some creature without shoes. Major Waddell was puzzled as to what species of animal with humanlike feet could have made the crossing of this isolated pass in that extreme temperature in bare feet. It had to be assumed that it was a Hindu fakir or some type of religious fanatic who had made the journey.

Following that original record, a number of rumors began to come out of the area about the presence of strange animals. One of them occurred in 1902 when Britain was stringing a telegraph line from Lhasa, the capital of Tibet, to the town of Darjeeling, in India. One night in an area near the spot where Tibet and Sikkim shared the border, a number of workers failed to return to their camp. A group of soldiers sent out could find no sign of them, but they found a peculiar animal lurking under a ledge of rock and fast asleep. The Indian soldiers immediately shot and killed the animal. On examination, they found it looked more like a human than an animal. It had a thick coat of hair. One of the reports claimed that the animal was ten feet high with a hairless face and long yellow fangs. The carcass was reported to have been packed in ice and shipped to Sir Charles Bell, a British political officer stationed in Sikkim.

In 1951 a forestry officer, J. R. O. Gent, who was located in Darjeeling, claimed to have seen a strange creature, but he was not quite sure whether it was a large monkey or an ape. It lived in the mountains and descended into the valleys only when the weather was very cold. He described it as being covered with long hair and with a hairy face. The hair was yellowish brown in color. The creature was only about four feet high, but it had very long feet, and according to Gent, its feet pointed backward when the animal was moving. This suggested that the animal walked on its knees and shins. Gent could not identify the animal, but he did conclude it was not a Nepal langur, which is a type of monkey. In 1907 A Russian zoologist, Vladimir A. Khakhlov, was studying reports of similar activities in Eurasia. The legends concerning these strange creatures had come from an area over thousands of miles in length, covering the whole of the Himalayan range, including reports from Tibet, Nepal, Sikkim, Assam, and Bhutan. By 1920 the English-speaking public still had very little knowledge of these legends and rumors. But in that year an event occurred which made these creatures front-page news. An expedition was launched to explore the northern face of Mount Everest. At 17,000 feet the leader of the expedition, Lieutenant Colonel Sir C. K. Howard-Bury, saw through his binoculars on one of the passes some dark forms on a snowfield above him. These forms were moving about, and the expedition set out to reach the snowfield where they could be seen. When they arrived, the animals had departed, but had left behind a large number of footprints which were described by Lieutenant Colonel Howard-Bury as being three times the size of normal human footprints. These creatures were seen by all the members of the reconnaissance party.

The Sherpa with the expedition believed the creatures that they had seen were human in form and they described them as *Metoh-Kangmi*. Howard-Bury sent a report to Katmandu, the capital of Nepal, and asked that his report be telegraphed to India. In the telegraphing of the report a mistake occurred. The word *Kangmi* appears to mean "snow creature," and *Metoh* was mistranscribed by the telegrapher as *Metch*. Actually, *Kangmi* and *Metoh* would have meant "snow creature." The recipients did not understand what *Metch-Kangmi* might mean, and they asked a columnist for the Calcutta *Statesman,* Henry Newman, to translate for them. He translated this name as meaning "Abominable Snowman." The press was delighted with this expression, and it made headlines first in the British press and then all over the world.

In November, 1921, William Knight, an Englishman, told the London *Times* of an encounter he had in a town called Gangtok when he was returning from Tibet. He was on his horse and had stopped to give the horse a breather when he looked around and saw a figure standing about twenty yards away. The animallike creature was not quite six feet in height, wore no clothes, had a pale yellow color and not much hair on its face, but a lot of hair on its head and very strong muscular development. Knight believed that he had seen an Abominable Snowman.

In 1925 A. N. Tombazi described his Snowman encounter in the Himalayas in the region of Sikkim. From a distance of two or three hundred yards he witnessed a figure that was shaped like a human. This creature was moving around in the upright position, occasionally uprooting rhododendron bushes. It appeared to have either dark skin or dark fur, and it was naked. After Tombazi looked at it for a minute or two, it moved away and was lost in the surrounding scrub. There had not been time to take photographs.

The Abominable Snowman has been described by some people as a sadhu, an Indian hermit who lives at high altitudes even up to 15,000 feet in the Himalayas. It seems very likely that some of the sightings that have been described may very well have been one of these people. Perhaps even some of the tracks that have been observed may have been human tracks.

In the 1930's there was an epidemic of people finding strange tracks of bare feet in the snow strange tracks of bare feet in the snow. Commander E. B. Beauman saw tracks at 14,000 feet. Eric Shipton, when returning from his climb on Everest, saw similar tracks at 16,000 feet, claiming that they looked rather like the tracks of a young elephant. The length of the stride suggested that they were made by a creature that walked erect. Not long after Shipton's observation Ronald Kaulback also saw, at 16,000 feet, tracks that looked as though they had been made by bare human feet. This occurred in the area southeast of Tibet. Kaulback did not accept that they had been made by man. It has been suggested that these tracks may have been made by bears, but since there was no evidence of bears in this region, Kaulback thought that the tracks had been made by a snow leopard. The porters believed they had been made by a Mountain Man. They described him as having white skin and long hair dropping from the shoulders and also on the arms. In addition, the creature had a long head of hair.

The Sherpa call these Mountain Men, or Abominable Snowmen, *Yetis*. The unusual tracks were made, they claimed, because the *Yeti* has additional toes and when it walks, it has its feet twisted from back to front. In 1937 Sir John Hunt, the famous Everest explorer, found strange footprints in the snow in the mountains. In 1938 an H. W. Tilman found footprints in the same place that Hunt had discovered them. These footprints were at 19,000 feet. Sen Tensing, a Sherpa who was with them, claimed that the tracks were those of *Yetis*. He said that he and other Sherpa had previously seen the creatures only twenty-five yards away and that they appeared to be hairy, manlike creatures about 5 feet 6 inches in height with rather pointed heads. Tensing was asked about these creatures when he attended a reception at the British embassy in Katmandu, and he made the very interesting observation that when the animal moved slowly, it remained upright, but if it had to move anywhere in a hurry, it ran on all fours.

Prince Peter of Greece is an anthropologist who also became interested in the Abominable Snowman and wrote a letter to the Indian newspaper *The Statesman,* describing some information that he had acquired which was relative to the creature. In one of the valleys in Sikkim an individual described as a Snowman had been coming up the valley at night and drinking from a cistern. The villagers grew nervous at this mysterious stranger and prepared a bucket of liquor for it. The animal drank the bucketful of liquor and passed out. It was discovered in the morning and tied to a pole. Later on, when it became sober, its strength was such that it was able to break the ropes that bound it to the pole and get away.

Dr. E. Wyss-Dunant, a Swiss who led an expedition in 1952 to Everest, also discovered footprints identical with those that Shipton had seen and photographed. He found at an altitude of 19,000 feet evidence that a group of *Yetis* could have possibly followed them. What they saw were tracks crossing their own. Wyss-Dunant claimed that the big toe was separated from the four other toes rather like the footprint of an ape and, further, that the tracks had not been made by creatures standing on two legs but by one individual on four legs. He believed it to be shaped similar to a bear and estimated the creature's weight to be somewhere between 168 and 217 pounds.

Sir John Hunt, associated with a successful Everest expedition in 1953, visited a monastery at the town of Thyangboche and asked the monks whether they had ever seen any *Yetis.* They claimed they had seen one four years before and that it frolicked in the snow only 200 yards away from the monastery. It appeared to be five feet in height and had grayish-brown hair. It moved about by alternately walking on its hind legs and then crawling around on all fours.

In 1954 an expedition was financed by the *Daily Mail* of London and set out for the Himalayas to find and photograph the Abominable Snowman. This time the expedition included not only people of mountaineering experience but also scientists. Included in the expedition were Dr. A. Biswas, an Indian from Calcutta, and also an English explorer, Gerald Russell. The expedition failed to find any Abominable Snowmen, but it was the first of the expeditions to treat the matter with serious intent.

In 1955 an expedition from the Royal Air Force reported they had seen tracks of the Abominable Snowmen. John Keel in 1958 claimed that he had not only followed the tracks of an Abominable Snowman but that he had actually sighted one. In 1957 the Texas multimillionaire Tom Slick organized an expedition to the Himalayas for the purpose of finding the Abominable Snowman. This brought together Gerald Russell and Peter Byrne and Bryan Byrne. Peter is now a leading investigator of the American Bigfoot; he is also an experienced animal collector.

At about the same time there were a number of reports turning up in the Soviet Union from people, especially in the Mongolian People's Republic, who claimed they had seen an Abominable Snowman or some such creature. Some reports even came from China. The Russians in 1958 sent an expedition to the Caucasus and another in the area of Mount Everest to a region called the Pamirs, a location the Russians believed could be the breeding grounds of the Abominable Snowman. Authorities of the Soviet Union and from China both arrived at the conclusion that the Abominable Snowman was a hominid, a manlike creature, rather than a pongid, which is apelike.

According to journalist Robert Chapman, writing in the London *Express,* the Russians still show great interest in the Abominable Snowman, particularly one that is believed to lurk in remote areas of the Caucasus. Professor Dmitri Bayanof, an authority on hominids, predicts that an expedition which is now investigating will probably extend its observations well into the future. The Russians believe that these strange hominids exist and call them *Almas.* Although there have been no direct sightings, the expedition has extended its search because of the more than 300 reports of sightings by various people and the discoveries of footprints in the ground. Even small stores of food have been found stored in forest glades. These cached supplies included pumpkins, potatoes, and corncobs. The people who have been reporting these sightings come from a variety of walks of life: Caucasian shepherds, tea planters, and dairymaids. Some of them reported seeing footprints, and others actually claimed to have had direct encounters with the *Almas.* For their descriptions the Darwin Museum in Moscow has constructed an "identikit," a complete likeness which provides a clue to the unclassified creature, be it *Alma,* , Snowman, or *Yeti.*

The Russians maintain that when this hominid is full grown,

it is over six feet tall and covered with thick, dark hair. Its head is small and egg-shaped, and its brow is receding. It has a broad, flat nose, and its lower jaw protrudes and has prominent teeth. The hominid creature shambles along with a pigeon-toed gait. Its arms are long and dangling, and as it moves, it mutters to itself. Its eyes are reddish and slitlike. It has a belligerent appearance, but apparently takes fright very easily and has never been known to attack anybody. Both males and females have been seen. The expedition which was launched in 1974 was headed by Professor Jeanne Koffman. Professor Koffman, a French subject, went to Russia during the Second World War and became a Soviet citizen and is now a member of the Soviet Geographical Society. She and an assistant planned to stay in the mountains during the winter months, hoping they might make contact with *Almas.*

The first reports of *Almas* in the eastern area of the Soviet Union were made as long ago as 1870 by an officer in the czarist army.

Many people have said that because there have been so many sightings of footprints of the *Yeti,* it is very surprising that one has not yet been captured. Ivan Sanderson has pointed out that sightings of the *Yeti* have been claimed, especially on the Tibetan plateau, and that this is also the area where the giant panda has been hunted, and the history of the panda is rather like that of the *Yeti.* In 1869 a French missionary in the province of Szechwan first saw a panda, which he described as a bearlike creature. Father David in 1900 was able to obtain a dead specimen, but after that year these animals were not seen again until 1915. In 1929 the two sons of President Roosevelt, Theodore, Jr., and Kermit, were on a hunting expedition in this part of the world, and they managed to tree one panda and then shoot it out of the tree. In the early 1930's a museum expedition managed to shoot two or three more. In 1937 and 1938 Ruth Hargness captured two pandas, which she sent to the Brookfield Zoo in the United States. It is an interesting fact that the giant panda lives in the same general area and also at the same elevation as the *Yeti.* For more than seventy years the giant panda had been hunted, and not one had been captured alive, so it is not surprising that if the *Yeti* does exist, he has not yet been captured, particularly since he is probably much smarter than the giant panda and so has been able to keep out of reach of people who are anxious to get their hands on him. One of the monasteries in Tibet claimed to have the scalp of a *Yeti,* but when it was examined, it was found to be made of the skin from a shoulder of a Himalayan goat antelope. Many of the so-called *Yeti* footprints have also been explained as ordinary animal footprints which have lengthened by the melting snow.

In 1958 there were a number of reports of what appeared to be the equivalent of an Abominable Snowman in California. The American Indians of the Northwest states and British Columbia have old legends which tell of creatures like the *Yeti*, tall, hairy creatures walking in a bipedal fashion, which have been known in that part of the world for generations. The first record of a *Yeti* in North America, apart from the Indian legends, dates back to 1811. At this time a party of explorers, under the leadership of David Thompson, found some large footprints, eight inches wide and fourteen inches long; these were larger than a bear would have made. Thompson tried to get his Indian guides to follow these tracks, but they refused.

One account which attracted considerable attention in 1884 concerned a railway-construction crew that came across a creature, not unlike a gorilla in appearance, lying on the grass at the side of the railway tracks. He was captured and named Jocko and was held in captivity for some time. There is at least one person who claims to have seen this animal in the captive state and described it as not human, but more like a human than an animal. No one ever found out exactly what became of it. Sanderson lists a number of theories regarding this creature. One was that the animal was a hoax; second, that it was some sort of cross, although what kind of hybrid it could be he did not explain. It may possibly have been a throwback to an earlier period or conceivably someone who had been lost many years before, as a small boy, and had now grown up and managed to survive in the wilderness on his own. Other suppositions were that it could be a mental defective or glandular case from an institution or an ape that had escaped from a circus. The inhabitants in the area at the time simply thought it might have been a primitive human or some species of great ape as yet unidentified. The reports given of this New World Abominable Snowman suggest that the animal stood about eight feet in height and had a very short and hairy neck and that the body hair was a reddish-brown color.

These animals were said to stand upright and to remain upright as they walk. In this case they differed from bears and, of course, from great apes, both of which move about on all fours. There have been a variety of names for this creature of North-western North America. Some have called him Susquatch, which means "a wild man of the woods." In other parts of the country he is called by an Indian name, Omah, but generally this mysterious creature is known to the public as Bigfoot.

A new personality moving into the Bigfoot scene in more recent years is Peter Byrne, a well-known animal tracker and big-game hunter. At one time he used to squire American sportsmen visiting Nepal on tiger hunts. When he became aware that extermination of the tiger was not far off, he established a tiger sanctuary. He is also one of the founders of the International Wildlife Conservation Society in Washington, D.C. Byrne has been on several expeditions in the Himalayas hunting *Yeti* and has now become interested in the Susquatch and has set up a camp in The Dalles, Oregon.

Byrne's camp has an observatory which looks over the region where there have been a number of reports of Susquatches, and he has recently started a small newspaper called the *Bigfoot News*. Byrne believes that if there are any Susquatches, they would number no more than 100 or 200, spread out over an area of thousands of square miles, which is very low population density. That means that the chances of seeing them, particularly since they restrict themselves to isolated areas away from humans, would be small.

A few years ago a rancher, Roger Patterson, came out of the wilderness area near Bluff Creek, California, claiming that he had not only seen Bigfoot but had photographed it. The film has been shown to a number of scientific organizations and groups, and it has been shown at the Yerkes Center also. It is not of very good quality and was taken more than 100 feet away. In the film a hairy, bipedal animal with a walk that looks suspiciously humanlike can be seen walking, stopping, and looking. It appears to be around seven feet in height, but it left footprints which were relatively small for such a big animal— only fourteen and a half inches long. The creature had an obvious crest to its head like a male gorilla. It also appeared to have some folds in the chest region which looked rather like fur-covered breasts, so it had the characteristics of a male and a female gorilla in one animal. It is unusual for the breasts to be completely covered with hair; they tend to be virtually free of hair in nearly all animals. Dr Donald Grieve of the Royal Free Hospital of the University of London has made a frame-by-frame analysis of the film, paying particular attention to the movements of the leg and the length of the stride and the time of the swing of the leg. He was not able to come to a clear decision on the creature. The Russians also made an analysis of the film at the Moscow Museum and seem to view it more favorably.

Some Russian scientists appear to be convinced that the film itself and some of the plaster casts of the footprints are both genuine. Professor Bayanof says of the film:

> We have established five correlations between the footprints and the creature seen walking in the film. All five are distant from, or totally non-existent in sapiens characteristics. . . .
> Thus American practice and Russian theory in hominology converged to produce what we think is a very impressive scientific blend. . . . The upshot of all this is that there is evidence which makes the creature's photographic appearance and movements available to everybody's eyes.

The Russians believe that their *Almas,* the Abominable Snowman, or *Yeti,* and the American Bigfoot are actually the remains of a race of dawn men who managed somehow or other to survive into the twentieth century.

An American movie studio which specializes in animation, after seeing the Patterson film, concluded that the subject was an animal and not a man in a fur suit. The English anatomist John Napier also saw the film and believes that it is inconclusive. He feels the same about many of the eyewitness reports.

If the Susquatch does exist, Napier believes it could be a *Paranthropus* which was an evolutionary by-product that diverted from the main course of human evolution some 2,000,000 years ago. It is hard to be sure of the size of the creature if it exists; just as many people in the past described the gorilla making a bluffing rush at them as being of immense size, so the frightening aspect of a Susquatch might also magnify it in size. If the footprints which are alleged to have been made by it are genuine, however, they do suggest that the animal is considerably larger than man.

Though there are numbers of people today described as weekend warriors that go hunting for a Susquatch, equipped with all kinds of guns with the intention of shooting the animal if they see one, Peter Byrne considers this inhumane. Instead he plans to shoot the animal with a tranquilizer gun so that it can be captured and studied.

There have been claims that creatures like Bigfoot are present in other parts of America, for example, in the south Florida swamplands. In Atlanta, Georgia, a Bigfoot research and investigation team has been organized by Bill Allen. This group is dedicated to the capture of Bigfoot and other hairy creatures. Last November they went to the south Florida swamplands hunting for a creature described as gorillalike which had been named the "skunk-ape" of the Everglades. A few months before the arrival of the expedition there had been a report of a motorist hitting and injuring one. At least 100 sightings of these creatures have been reported in Florida over the last three years. Together with Gale Morris and Jim Lyles, the other members of the search team, Allen claims to have found evidence that probably at least five Bigfoot-like creatures are roaming in the south Florida swamplands. There was evidence of a raid on their campsite, and the team made plaster casts of some of the footprints, which they claimed were left by the animals. Allen and his associates claim these creatures are apelike in appearance, but more like humans in intelligence. The Florida swamp harbors two species of Bigfoot. One is a small animal which is gentle and has five toes, but there is a larger one, a three-toed creature with a mean disposition. Allen and his group maintain that the *Yeti* from the Himalayas, the Bigfoot, and the skunk-ape of Florida are all related. There is, of course, a size difference and a difference in fur. According to Allen, the Snowmen are covered with a dirty white fur, while the Bigfoot and skunk-ape are covered with a black, ginger-colored, or red-hued fur. In their last expedition to south Florida Allen claims that members of his team managed to sight several skunk-apes, but were not able to get satisfactory photographs. One photograph was taken, but it showed only the top of the head.

On another expedition a Bigfoot or skunk-ape threw seashells at one of the members of the team, and when they fired their guns in the air to try to scare him and other Bigfeet off, the creatures became defiant and shook the trees surrounding the campsite. Allen contends that these creatures are very intelligent and this explains why they are able to elude would-be captors. On one occasion, it is said, one of the creatures they were stalking stole a hat and a life jacket out of their boat. Allen claims that another emptied one of their bottles of scotch; whether Bigfoot had a big hangover or not Allen didn't say. It has been suggested that these creatures have strong family ties and react very protectively to possible danger or attempted capture of any member of the family. In the past some of these animals have been reported killed, but no one seems to have recovered the carcasses.

A West Coast commercial fisherman, Dick Grover, devotes a good deal of his spare time in investigating reports of the sightings of Susquatch or even camping in areas where Susquatch is said to have been seen. He now has an organization which he calls Project Discovery. Grover claims he knows a family, whom he wishes to keep anonymous for the time being, who were in a wooded camping site in the northwest of the country and eventually fled from the site because they were terrorized by a group of screaming hairy creatures, who apparently were angry and threw boulders at them. Grover's investigators do not wish to kill a Bigfoot or to capture one, but would like some "extremely hard-core evidence; something you can put your finger on. We hope one day to be able to come up with close-up photographs, including close full-face views."

Grover maintains that his group is looking for an animal which is "highly intelligent, but more animal than human, living in heavily wooded areas, but not necessarily remote, uninhabited mountain country." The animals that they are looking for, he said, would probably be eight feet tall and weigh as much as 800 pounds. He believes there are not more than 150 to 200 of these animals in existence and that they are probably in the state of Washington.

Peter Byrne's object in establishing the Bigfoot Information Center in the The Dalles, Oregon, was to bring to people all over the United States factual and educational information on the Bigfoot phenomenon. His four-page sheet, the *Bigfoot News,* lists all the latest information about the Bigfoot, including the latest reports of sightings of the animal or of tracks or of anything else that is relevant to this creature. It will also sponsor teams of field associates and scientists that will actually launch expeditions to find these creatures or study them if they can be obtained.

In a letter of July 26, 1974, Peter Byrne mentioned that at one time he was very skeptical of the existence of the Bigfoot and that he is convinced that there is a small group of them in the mountains around the area where he has established his information center. He believes he will come to grips with them in due course. He feels there is simply too much evidence and too many sightings by many reliable, honest, unimaginative people for them not to be true. He pointed out that he has been searching in the area now for three and a half years, but feels the goal is worthwhile and plans to stay with the problem until he solves it. He was good enough to send a photograph of a sixteen-inch footprint that he found in Bluff Creek in north California in 1960, which is reproduced in these pages. Some recent news of Bigfoot include the following:

In the Carcel Rock area in British Columbia there was a report from a man and his wife who were driving along a lonely road in the area one night and saw a Bigfoot standing in the road. They slowed down, but continued on and did not stop. They estimated that the creature was about six and a half feet in height, massively built, and covered with hair which appeared, as far as they could tell at night, to be either dark brown or black. The animal did not move while they passed it. Though they did not come back and look for prints, they did report their experience to the Bigfoot Information Center.

In the Little Applegate River area in South Oregon, Don Brown of Medford, Oregon, came across some footprints. Brown and his two companions observed prints extending for 400 yards, Each print, about fourteen inches long, was quite clear and had five toes.

On November 12, 1974, a skipper from a logging boom tugboat and one of his crew said that they saw a huge manlike figure on the rocky shore in the Queen Charlotte Straits. It stood there for three or four minutes, then it turned and walked into the dense brush.

In the state of Oregon on July 12, 1974, Jack Cochran was operating a logging crane and saw a figure which he took to be a human watching him from the edge of the cutting which had already been cleared of timber. Although at first thinking it was a man, he remembered there were only two of them in the area and that the man who was working with him was in sight some distance behind him. The creature that he observed was massively built, particularly in the shoulders. It was covered with hair which was dark in color and walked upright. It walked from the clearing into the area of dense timber, and Cochran was so intrigued that he got out of the cab of the crane to observe it more closely. His companion, unfortunately, did not see it. The next day, however, at approximately the same time, two lumbermen working in the same area took a work break during which they walked to the edge of the forest. Immediately in front of them, on the fringes of the forest, they saw a very large, heavily built figure covered with dark hair that had been squatting in the undergrowth. The figure jumped to its feet and moved quickly away. One of the men, not sure what it was, ran after it to try to get a better look at it. It outpaced him without much difficulty and disappeared into dense vegetation. When they first saw the creature, it was only thirty feet away from them. When the area was examined afterward, there was evidence that there had been a large and heavy bipedal creature there which had left footprints. Investigators from the information center spent ten days in the area making an intensive search, but were not able to find any further trace of this Bigfoot. All the information available, however, according to the *Bigfoot News*, seems to support the truth of this story.

Among the mountains of Oregon is one located on the edge of the Clackamas River drainage, called Round Mountain (4,668 feet), and near it is Tarzan Springs, which is the locale of a Bigfoot sighting. Glenn Thomas, a timber worker, claims that in 1967 he saw three Bigfeet in this area digging up rodents and eating them. One morning Thomas walked down a logging road near one of the stony ridges. He heard a noise of rocks falling and stopped and saw three large figures digging into a rock pile. The creatures did not see him because he was screened by a clump of trees, but he observed that the group consisted of a large male, a smaller female, and a young "animal." They could be seen lifting rocks which may have weighed as much as 200 pounds. They dug down at least six or seven feet, then the male creature reached into the hole and took out a nest of rodents, and all three of them reached for the rodents and ate them. The creatures then walked away and were not seen again. The information center investigators went to the area in 1972 and again in 1973. The rodents of that area were found to be woodchucks, or marmots. In October, when the hairy creatures are supposed to have been seen, marmots are in deep hibernation and can be handled at that period without their waking up. They found thirty additional holes apart from the one that Glenn Thomas claimed to have seen the animals digging. Around these holes were rocks which did, in fact, weigh 250 to 300 pounds.

In 1974 there were two expeditions chasing after Bigfoot. Russ Kinne, who wrote an article on Bigfoot for the *Smithsonian* magazine during the summer of 1974, joined Peter Byrne and Fran Townsend Dickinson to search for Bigfoot. They carried an Explorers Club flag with them. They went to British Columbia, where they explored near Williams Lake and then over to the coast via Belle Coola. They searched the areas both from the air and from the ground. The areas included Bute and Sus-quatch Pass. They also searched Night Inlet and the Homathko River and Southgate River areas. They landed their plane on many old logging strips and on some of the logging roads in the mountains and also in a number of fields. Territory that they were not able to cover satisfactorily from the air they walked through on foot. They did some extensive hiking with back-packs. They were not successful in seeing any Bigfeet, but they did obtain what was claimed to be the leg bones of a Bigfoot and photographed them. These photographs are being studied at the moment by the American Museum of Natural History.

The absence of the remains of any skeletons or parts of skele-ton of Bigfoot is a question that is often raised. The same ques-tion also occurred to Fred Blaine of Baker, Oregon, a rancher in that area for forty years. He wanted to know why, with so much talk about Bigfeet, and he had heard it when he was a boy more than forty years ago, no bones or skulls had ever been found. The *Bigfoot News* answers this question by explaining that one of the reasons that very little bone material is found is that the coastal ranges in quite a number of places have a wet acid soil. This is not conducive to the preservation of bone. They point out also that nature has a disposal system: A lot of animals and birds eat the bodies of other animals when they die. The black bear, apparently, is a principal scavenger in the woods and would, of course, very likely chew up many of the bones. Further, if you ask people in the area how many of them have ever seen or found a full skeleton of a deer or have ever seen the complete skeleton of a bear or a cougar, you find that few of them ever have. In that case it would be possible to un-derstand why practically no skeletal remains of the Bigfoot have been found. This lends still more interest to the photo-graphs of the alleged bones of the Bigfoot taken by the Russ Kinne and Peter Byrne expedition, and it is hoped that the American Museum of Natural History will have something of interest to say about them.

Another expedition in 1974 was one entitled the National Wildlife Federation Expedition. No members of the founda-tion actually took part in it, but it allowed its organization to be used as a channel for funds and also allowed its name to be used in connection with the expedition. The expedition was, in fact, backed principally by Mrs. Louisa Carpenter of Fort Lau-derdale, Florida, who contributed $40,000, and Mr. Caruth Byrd, who contributed $5,000. A Boston-based movie company also contributed $5,000 for the rights to make a movie of the expedition. The investigative group consisted of three scientists, a carpenter, an attorney, an orthodontist and his wife and two children, three young women, and an ex-computer technician, who was the leader of the expedition. The group was based at The Dalles in the beginning, then later they moved to the town of Cougar in Washington. Members of the group stayed on for varying periods of time. By July the sponsors ceased to support the project any longer, and the National Wildlife Federation withdrew its name from the expedition. The expedition did not make any finds, and they were not able to report anything of value at the conclusion.

Whereas the big hairy creatures seem indigenous to the mountainous areas of the world, many other countries have laid claim to at least one hairy monster. In the forests of Indonesia there is believed to be a manlike monster called an *Orang pendek;* from South America have come stories from the days of the Jesuit missionaries of legendary strange creatures— the *Mapinguary* and *Didi* among them. There was also a legendary monster in the Caucasus which turned out to be of a totally different kind. A newspaper reported around the year 1840 that some people in a Caucasian village had chased and killed a hairy monster. A few years ago scientists decided to investigate this report. Some of the elderly villagers in the area where this occurred recalled what their parents and grandparents had told them of this event and even remembered where the monster had been killed and buried. When the creature was exhumed, the skeleton turned out to be simply that of a normal human female. Apparently what happened was that some poor hairy woman was set on by the villagers, who believed that she was a witch or monster, and had been killed and buried.

In 1934 Ralph von Koenigswald, a Dutch paleontologist who worked extensively on the Java man *(Homo erectus),* was walking through the streets of Hong Kong and went into a Chinese druggist's shop. He was curious about Chinese drugstores, for in these stores paleontologists had been able to find all kinds of relics of extinct animals. What he came into the drugstore to look for specifically were shells or dried or stuffed animals or insects and possibly pieces of fossil bones. There was a jar of teeth on the counter, and while he was waiting to be served, he removed a handful and spread them out on the counter top and amused himself trying to determine which species of animal the teeth belonged to. Then, suddenly, he stopped, then slowly and carefully picked up one tooth which had particularly attracted his attention. It happened to be the third lower molar, but he had never seen any tooth of that size before. It was six times bigger than the third molar or any molar of a human. He immediately queried the chemist as to where the tooth had come from, but the druggist had no idea of its origin. It had been in the shop for a long time and may even have been brought into the store originally by his father or his grandfather. It may have been collected even further back in his ancestral history. Such "dragon's teeth" were often found by his relatives while digging in the fields.

This find was enough to send von Koenigswald into every drugstore in Hong Kong and to every drugstore he came across in China. Eventually in Canton he came across another tooth which was identical with the one that he had found in Hong Kong, but this time it was an upper, rather than a lower molar. He continued his search and after five years he found a third molar, a second one from the lower jaw, in a good state of preservation, better than either of the other two. The crown was intact, and the root was still there. All these teeth were about the same size and were twice the size of the molar teeth of an adult male gorilla. If the animal to which these teeth belonged was proportionate in size to the size of his teeth, he would have stood about twelve feet high. It was impossible to tell from the teeth whether they had actually come from a large ape or from a large hominid. Professor Elwyn Simons of Yale University thinks that the animal belonging to the teeth von Koenigswald found actually stood about eight feet high and may have weighed as much as 600 pounds, or about the same weight as a large male gorilla. Coincidentally, it was the study of "dragon's teeth" in Chinese drugstores by Professor Davidson Black, a medical missionary working in China, which stimulated him to look for remains of early man in the area around Peking.

Von Koenigswald showed his collection to another paleontologist, Franz Weidenreich. Weidenreich inspected the teeth and decided they were not those of a giant apelike creature, but of a giant man. He suggested it be named *Gigantanthropus*, for "giant man," instead of *Gigantopithecus*, the name given to it earlier by von Koenigswald, which meant "giant ape."

By 1952 von Koenigswald had in his possession eight teeth all found at Chinese drugstores and considered that these teeth were definitely human in nature and that they were around 2,000,000 years old.

In the 1950's Dr. Pei Wen-chung, the Chinese paleontologist, began excavating in the mountains in Kwangsi Province in South China and discovered three complete jawbones of *Gigantopithecus*. By 1955 Dr. Pei had forty-seven *Gigantopithecus* teeth and discovered three more in 1956. In the village of Chang Tsao, at about this time, a Chinese farmer called Chin Hsiu-huai, who was collecting fertilizer from a mountain cave, also in Kwangsi, found a large jaw with some teeth attached to it, and he gave it to the Chinese state. A group of scientists from the Academia Sinica began excavations in this area and by 1958 had found two more lower jaws; the third one was enormous and the first one relatively much smaller. There were no signs of stone tools in the caves in the excavations from which these lower jaws had been found. Certain anatomical features of these jaws indicated that the animal was completely different from the gorilla, and according to Dr. Pei, the wear on the teeth indicated that the animal was not a vegetarian, but probably ate a mixed diet which included meat. It was apparent that the front teeth, which are called incisors, were fairly small, again suggesting that these animals were not apes. Simons believes the wear on the teeth of *Gigantopithecus* indicates that there may have been increased consumption of grains by these animals which required considerable chewing, resulting in wearing of the surface of the teeth. Simons also believes that *Gigantopithecus* was not much of a meat eater, if at all.

The remains of *Gigantopithecus,* in China, probably date from about 500,000 to 1,000,000 years ago. We are suggesting that the Abominable Snowman of Tibet and the Susquatch of North America may be specimens of *Gigantopithecus* which may have survived to the present day. This does not appear to be out of the realm of possibility. An example of how evolutionary relics can remain in isolated areas for a long period of time is evidenced by the fossil man *Homo erectus*. Here was really the first true man, having a brain capacity approaching 1,000 cc, which places it in the human range. It was also an erect animal, and, as pointed out earlier, one form of *H. erectus* was found in China, another in Java. There is now evidence that some 1,000,000 years ago *H. erectus* probably got to Australia, where he found no other subhuman or other hominoid forms or even apelike forms to compete with because Australia had been cut off from the rest of the world for many millions of years. In this isolated situation *H. erectus* lived in modest numbers until some 50,000-odd years ago, when the predecessors of the Australian aborigines came to the island on rafts or in dugout canoes. Apparently there was some racial mixing between the two, but there is now evidence that *H. erectus* lived in Australia until only a few thousand years ago. It does not seem impossible, therefore, that *Gigantopithecus* has survived in remote areas in Asia and North America and might be the animal that is responsible for the *Yeti* and Bigfoot sightings. It may indeed have survived to the present day in very isolated areas.

There is no certainty as to whether *Gigantopithecus* was quadrupedal or erect, but it seems likely that he was not truly erect, although he may have evolved more in that direction over a 1,000,000-year period. The sightings in the Himalayas where the animals were observed to be standing up on their legs at first and then dropped onto all fours to move away seem to provide additional argument in favor of the Snowman as a relic of another age.

It is of interest that Simons and his group have also found the remains of a creature in India which showed many similarities to the *Gigantopithecus* and was, in fact, classified by them as such. *Gigantopithecus* remains have been found in China, and he could very easily have been at one time in Tibet, India, and many other parts of the Himalayas, but how could' *Gigantopithecus* get to the North American continent, which has no primitive man at the level of *H. erectus* or earlier? There is an answer to this puzzle. It is that from the province of Szechwan there is a mountain causeway through Manchuria and up into the eastern part of Siberia. This causeway is covered in montane forest. This same forest can also be found on the other side of the Bering Strait in Alaska and passes down the western side of the North American continent all the way down to Tierra del Fuego, which, as everybody knows, is at the bottom of South America. We know that during the periods when the ice of the North Pole advanced and retreated, there were times when there was a complete sheet of ice all the way across the Bering Strait. A number of Siberian animals managed to get across to the northern part of the New World. Included among these are, of course, the elk, the moose, and the brown bear, and there is a good deal of evidence that the American Indian and also the Eskimo from Siberia took the same route. Therefore, there is no reason why *Gigantopithecus* could not have earlier come up the mountain causeway and crossed the Bering Strait, either over the ice or over a land bridge, and got into the montane forests of America by the same method. In fact, you can probably walk across the ice of the Bering Strait even now in the wintertime. So perhaps the *Gigantopithecus* is the Bigfoot of the American continent and perhaps he is also the *Yeti* of the Himalayas. Only the discovery of an actual animal and its thorough scientific examination can provide the answer.

With the legends and superstitions of Bigfoot, *Yeti*, and Susquatch, we have come full circle. Haunted by the same myths and fantasies, the gorilla has finally emerged, though still not safe from the slings and arrows of the prejudiced and the blindly ignorant, as a gentle vegetarian giant. But out of the same fabric of fantasy and illogic, out of the same darkness of superstition, come the new monsters—the hairy wild ones. If the *Yeti* and the Snowman are indeed real, will they likewise suffer the early fate of the gorilla? How long will it be before science can lay a restraining hand on the gun of the hunter?

Footprint of Bigfoot? (Courtesy of Peter Byrne.)

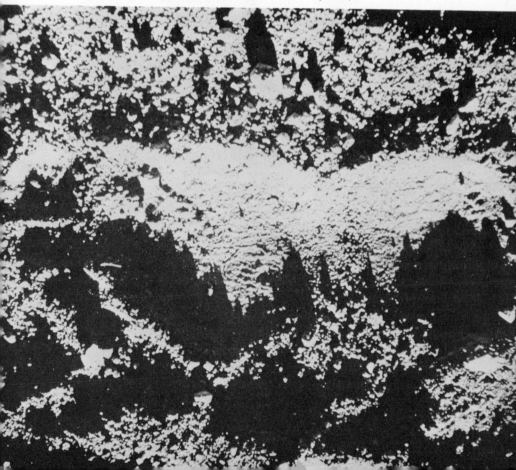

Still photograph from Patterson film of Bigfoot. (Courtesy of Peter Byrne.)

Bibliography

AKELEY, CARL E. "Gorillas, Real and Mythical," *Natural History,* Vol. 23 (1923).

AKELEY, MARY L. J. "Africa's Great National Park," *Natural History,* Vol. 29 (1929).

ANGUS, S. "Water-Contact Behavior of Chimpanzees," *Folia Primatologia,* Vol. 14 (1971).

ASAI, RIKIZO. *Method of Training Gorillas.* Nagoya, Japan.

ASHTON, E. H. "Age Changes in Some Bodily Dimensions of Apes," *Proceedings of the Zoological Society of London,* Vol. 124 (1954).

BALDWIN, L. A., and GEZA TELEKI. "Field Research on Chimpanzees and Gorillas," *Primates,* Vol. 14 (1973).

"Baby Gorilla, Bamboo," *Bulletin of the Zoological Society of Philadelphia* Vol. 5 (1930).

BAUMGARTEL, M.W. "The Muhavura Gorillas," *Primates,* Vol. 3 (1961).

BENCHLEY, BELLE J. "Gorillas Again," *Zoonooz,* October, 1949.

BOLWIG, NIELS. "Facial Expression in Primates," *Behavior,* Vol. 22 (1964).

———. "A Study of the Nests Built by Mountain Gorillas and Chimpanzees," *South African Journal of Science,* Vol. 55 (1959).

BOULENGER, E. G. *Apes and Monkeys.* London: George Harrop & Co., 1936.

BOURNE, G. H. *The Ape People.* New York: Putnam, 1971.

———. *Primate Odyssey.* New York: Putnam, 1974.

BURTON, MAURICE. "The Gentle Gorilla," *Illustrated London News,* Vol. 27 (1966).

CARPENTER, C. R. "An Observational Study of Two Captive Mountain Gorillas," *Biology,* Vol. 9 (1937).

CARTER, F. S. *Comparison of Baby Gorillas with Human Infants.* Jersey: Jersey Wildlife Preservation Trust, 1973.

CARTMILL, MATT. "Rethinking Primate Origins," *Science,* Vol. 184 (1974).

Cassell's Natural History, Vol. 1, 1877.

"Challenging Kenyapithecus," *Science News,* Vol. 96 (1969).

CHANCE, M., and JOLLY C. *Social Groups of Monkeys, Apes and Man.* New York: E. P. Dutton, 1970.

303

CHARLESWORTH, E. "Remarks on the Structure of the Gorrilla," *Journal and Transactions of the Victoria Institute,* 1887.

CLARK, W. E. LeGros. *The Antecedents of Man.* Chicago: Quadrangle Books, 1960.

———. "The Contrasting Morphology Found in the Wrist Joints of Semi-Brachiating Monkeys and Brachiating Apes," *Folia Primatologia,* Vol. 16, (1971).

CORDIER, CHARLES. *"Les Gorilles,"* Zoo, 1960.

COUPIN, F. *"La Croissance Chez les Anthropoides et Chez l'Homme,"* Revue Scientifique, Vol. 66 (1928).

CUNNINGHAM, ALYSE. "A Gorilla's Life in Civilization," *Zoological Society Bulletin,* Vol. 24 (1921).

DE BOUVEIGNES, O. *"Le Périple d'Hannon et les Gorilles,"* Zoologist, Vol. 8 (1950).

DOLLINOW, PHYLLIS. "The Non-human Primates: An overview," *Primate Patterns.* New York: Holt, Rinehart & Winston, 1972.

DONISTHORPE, J. "A Pilot Study on the Mountain Gorilla," *South African Journal of Science,* Vol. 54 (1958).

DU CHAILLU, PAUL B. *Equatorial Africa.* London: John Murray, 1861.

———. *Stories of the Gorilla Country.* New York: Harper & Brothers, 1868.

DUCKWORTH, W. L. "Note on a Foetus of *Gorilla savagei,"* Journal of Anatomy and Physiology, Vol. 33 (1898).

———. "Variations in Crania of *Gorilla savagei,"* Journal of Anatomy and Physiology, Vol. 30 (1895).

EIMERL, S., and I. DE VORE, *The Primates.* New York: Time/Life Books, Life Nature Library, 1967.

"Electromyography of Brachial Muscles in Pan Gorilla and Hominoid Evolution," *American Journal of Physical Anthropology,* Vol. 41 (1974).

ELWES, DOMINICK. *Lemuel Gulliver and Kivu, the Gorilla.* London: Lilliput, 1968.

FISCHER, GLORIA J. "The Formation of Learning Sets in Young Gorillas," *Journal of Comparative Physiology and Psychology,* Vol. 55 (1962).

FOSSEY, DIAN. "Vocalizations of the Mountain Gorilla," *Animal Behavior,* Vol. 20 (1972).

——— and ROBERT M. CAMPBELL. "Making Friends with Mountain Gorillas," *National Geographic,* Vol. 137 (1970).

———. "More Years with the Mountain Gorilla," *National Geographic,* Vol. 140 (1970).

GARNER, R. L. "Gorillas in Their Own Jungle," *Zoological Society Bulletin,* Vol.17 (1914).

GIFFORD, DENIS. *A Pictorial History of Horror Movies.* London: Hamlyn, 1973.

GOODALL, JANE. "My Life Among Wild Chimpanzees," *National Geographic,* August, 1963.

"Gorilla Born at Brookfield," *Brookfield Bison,* May 31, 1971.

"Gorillas at the Zoological Gardens," *The Field,* September 8, 1904.

"Gorillas in Their Own Jungle," *Animal Kingdom,* Vol. 17 (1914).

GOULD, S. J. "This View of Life: Sizing up Human Intelligence. Unpublished MS, 1973.

GREGORY, W. K. "In the Land of the Gorilla," *Evolution,* Vol. 3 (1931).

GREY, ANTHONY. "The Stationary Ark," *Illustrated London News,* October, 1974.

GROOM, A. F. G. "Squeezing out the Mountain Gorilla," *Oryx,* Vol. 12 (1973).

GROVES, COLIN P. *Gorillas.* New York: Arco Publishing Co., 1970.

HAGENBECK, CARL. *Beasts and Men,* trans. H. S. R. Elliot and A. G. Thacker. London: Longmans Green, 1909.

HANEGGER, R. E., and P. MENICHINI. "The Life of Primates: A Census of Captive Gorillas with Notes on Diet and Longevity," *Zoologischer Garten Zeitschrift Gesamter Tiergarten,* Vol. 26 (1962).

HORNADAY, W. T. "Gorillas, Past and Present," *Zoological Society Bulletin,* Vol. 18 (1915).

IMANISHI, K. "Gorilla: A Preliminary Survey in 1958," *Primates,* Vol. 3 (1961).

INAUE, M., and S. HAYAMA. "Histopathological Studies on Two Mountain Gorilla Specimens," *Primates,* Vol. 3 (1961).

JOLLY, ALISON. *The Evolution of Primate Behavior.* New York: Macmillan, 1972.

———. "Hour of Birth in Primates and Man," *Folia Primatologia,* Vol. 18 (1972).

———. *Man's Place Among the Mammals.* London: Edward Arnold, 1929

JONES, F. WOOD. *The Human Foot in Phylogeny Structure and Function as Seen in the Foot.* London: Baillière Tindall & Cox, 1944.

KAWAI, M., and H. MIZUHARA. "An Ecological Study of the Wild Mountain Gorilla," *Primates,* Vol. 3 (1961).

KEITER, MARY D., and LOUIS P. PICKETTE. "Report of Survey of Lowland Gorilla Births in Captivity." Seattle, Washington: unpublished MS.

KEITH, ARTHUR. "The Gorilla and Man as Contrasted Forms," *The Lancet,* Vol. 210 (1926).

KELLY, JOAN MORTON. "Bathing Beauties," *Zoonooz,* November, 1952.

———. "Gorilla Gourmets," *Zoonooz,* July, 1954.

KORTLANDT, A., and M. KOOIJ. "Protohominid Behavior in Primates," *Symposium of the Zoological Society of London,* 1963.

LANG, ERNST M. *Goma, the Gorilla Baby.* New York: Doubleday, 1963.

———. "Jambo, the Second Gorilla Born at Basel Zoo," *International Zoo Yearbook,* 1961.

———. "Where Have You Been? Out! What Were You Doing? Strolling with 6 Gorillas!" *Ciba Journal,* No. 39, 1966.

LEAKEY, L. S. B. "The Relationships of African Apes, Man and Old World Monkeys," *Proceedings of the National Academy of Sciences,* Vol. 67 (1970).

"Mammalia," "Primates," *Zoological Record:* 1870, 1872, 1874, 1875, 1878, 1879, 1881, 1882, 1883, 1887, 1889, 1891, 1892, 1893, 1894, 1895, 1896, 1898, 1899, 1900, 1901, 1902, 1903, 1905, 1906, 1907.

"Massa Is the Bengum Word for Gorilla," *Zoonooz,* November, 1968.

The Missing Link. New York: Time/Life Books, 1972.

NOBACHI, CHARLES V. "Growth of Infant Gorilla," *American Journal of Physical Anthropology,* Vol. 14 (1930).

NORTH, H. RINGLING, and ALDEN HATCH. *The Circus Kings: Our Ringling Family Story.* New York: Doubleday.

OAKLEY, KENNETH P. "Dating the Emergence of Man," *Advancement of Science,* Vol. 18 (1962).

ORCUTT, ADELLE. "Gorilla Notes," *Zoonooz,* January, March, April, May, November, 1950; May, July, August, 1951.

OSBORN, ROSALIE M. "Observations on the Behavior of the Mountain Gorilla," *The Primate Symposium of the Zoological Society of London,* No. 10 (1963).

PETERS, M., and D. PLOOG. "Communication Among Primates," *Annual Review of Physiology,* Vol. 35 (1973).

PFAFFMANN, C. "The Comparative Approach to Physiological

Psychology," *Annals of the New York Academy of Sciences*, Vol. 223 (1973).

PHILLIPS, TRACY. "The Gorilla and Man," *Man*, Vol. 28 (1928).

PLOOG, DETLEV. "Primates and Human Ethology," *Comparative Ecology and Behavior of Primates*, ed. R. P. Michael and J. H. Crook. New York: Academic Press, 1973.

"Present State of Our Knowledge of the Anatomy of the Primates," *British Medical Journal*, October 2, 1948.

RANDALL, F. E. "The Skeletal and Dental Development of the Gorilla," *Human Biology*, Vol. 16 (1944).

RAVEN, H. C. "Gorillas: The Greatest of All Apes," *Natural History*, Vol. 31 (1931).

READE, W. WINWOOD "The Habits of the Gorilla," *American Naturalist*, Vol. 1 (1868).

"Return to M'kubwa," *Zoo Sounds*, Vol. 9 (1973).

REYNOLDS, J. "Some Behaviour Comparisons Between the Chimpanzee and the Mountain Gorilla in the Wild," *American Anthropologist*, Vol. 67 (1965).

REYNOLDS, VERNON. *The Apes*. New York: E. P. Dutton, 1967.

RIGBY, LILLIAN R. *My Monkey Friends*. Bristol, Eng.: Arrowsmith Press, 1938.

RIOPELLE, A. J., and K. KUK. "Growing Up with Snowflake," *National Geographic*, Vol. 138 (1970).

———. and PAUL A. ZAHL. "Snowflake: The World's First White Gorilla," *National Geographic*, Vol. 131 (1967).

ROCK, MAXINE A. "It's Not Easy to Become a Good Gorilla Mother," *Smithsonian Magazine*, September, 1973.

ROSE, M. D. "Quadrupedalism in Primates," *Primates*, Vol. 14 (1973).

ROSEN, S. E. *Introduction to the Primates*. Englewood Cliffs, N.J.: Prentice-Hall, 1974.

RUMBAUGH, D. M. "First Year of Life: The Behavior and Growth of a Lowland Gorilla and Gibbon," *Zoonooz*, 1966.

——— and T. V. GILL. "The Learning Skills of Great Apes," *Journal of Human Evolution*, Vol. 2 (1973).

SABATER-PI, JORGE. "An Albino Lowland Gorilla from Río Muni, West Africa," *Folia Primotologia*, Vol. 7 (1967).

———. *Distribución Actual de los Gorilas de Ljamura en Río Muni*. Publicación del Servicio Municipal del Parque Zoologico de Barcelona, 1964.

———. "Rapport préliminaire sur l'Alimentation dans la Nature des Gorilles du Río Muni," *Mammalia*, Vol. 30 (1966).

———— and CLYDE JONES. *Comparative Ecology of Gorilla gorilla and Pan troglodytes in Río Muni.* Basel: S. Karger & Co., 1971.

————. "Notes on the Distribution and Ecology of the Higher Primates of Río Muni, West Africa," Tulane Studies in Zoology, Vol. 14 (1967).

SAKLINS, M. D. "The Social Life of Monkeys, Apes and Primitive Man," *Human Biology*, Vol. 31 (1959).

SANDERSON, IVAN. *The Monkey Kingdom.* New York: Hanover House, 1957.

SCHALLER, GEORGE B. *The Mountain Gorilla.* Chicago: University of Chicago Press, 1963.

————. *The Year of the Gorilla.* Chicago: University of Chicago Press, 1964.

———— and JOHN T. EMLEN. "Observations on the Ecology and Social Behavior of the Mountain Gorilla," *African Ecology and Human Evolution*, 1963.

SCHULTZ, ADOLPH H. *The Life of Primates.* London: Weidenfeld & Nicolson, 1969.

————. "Man as a Primate," *Scientific Monthly*, November, 1931.

SMITH, R. R. "Gorillas," *Journal of the Institute of Animal Technicians*, Vol. 25 (1972).

STEPHENSON, P. H. "The Evolution of Color Vision in the Primates," *Journal of Human Evolution*, Vol. 2 (1973).

STRAUS , W. L. "Fossil Evidence of the Evolution of the Erect, Bipedal Posture," *Clinical Orthopedics*, Vol. 25 (1912).

————. "The Riddle of Man's Ancestry," *Quantitative Review of Biology*, Vol. 24 (1949).

"Third-Generation Gorilla Birth," *American Association of Zoological Parks and Aquaria Bulletin*, Vol. 9 (1968).

THOMAS, W. D. "Observations on the Breeding in Captivity of a Pair of Lowland Gorillas," *Zoologica*, Vol. 43 (1958).

VON HOFFMAN, NICHOLAS. "Husbands and Other Lovers," *Washington Post*, April 1, 1975.

WASHBURN, S. L. "The Promise of Primatology," *American Journal of Physical Anthropology*, Vol. 34 (1973).

———— and R. S. HARDING. "Evolution of Primate Behavior," *The Neurosciences*, ed. Francis O. Schmitt. New York: Rockefeller University Press, 1970.

———— and E. R. McCOWN. "Evolution of Human Behavior," *Social Biology*, Vol. 19 (1972).

WILLOUGHBY, DAVID P. "The Gorilla: Largest Living Primate," *Scientific Monthly*, Vol. 70 (1950).

YERKES, ROBERT. "The Mind of the Gorilla," *Genetic Psychology Monographs.* Worcester, Mass.: Clark University, 1927.

———— and ADA W. YERKES. *The Great Apes.* New Haven: Yale University Press, 1929.

Zoo, No. 3, January, 1967; No. 3, January, 1970; No. 1, July, 1972.

Zoo News, July–August, 1974.

Zoonooz, February, 1951.

Index

311

312

318

QL
737
P96
B77

Bourne, Geoffrey Howard, 1909-
    The gentle giants : the gorilla story / Geoffrey H. Bourne and Maury Cohen. — New York : Putnam, [1975]
    319 p. : ill. ; 24 cm.

    Bibliography: p. 303-309.
    Includes index.
    ISBN 0-399-11528-5

    1. Gorillas.   I. Cohen, Maury, joint author.   II. Title.

310969 QL737.P96B67   1975   599'.884   75-25753
                                                               MARC

    Library of Congress   75